COMMERCIAL USE OF
BIODIVERSITY

COMMERCIAL USE OF
BIODIVERSITY

Resolving the Access and
Benefit Sharing Issues

Shivendu K. Srivastava

www.sagepublications.com
Los Angeles • London • New Delhi • Singapore • Washington DC

First published in 2016 by

 SAGE Publications India Pvt Ltd
B1/I-1 Mohan Cooperative Industrial Area
Mathura Road, New Delhi 110 044, India
www.sagepub.in

SAGE Publications Inc
2455 Teller Road
Thousand Oaks, California 91320, USA

SAGE Publications Ltd
1 Oliver's Yard, 55 City Road
London EC1Y 1SP, United Kingdom

SAGE Publications Asia-Pacific Pte Ltd
3 Church Street
#10-04 Samsung Hub
Singapore 049483

Published by Vivek Mehra for SAGE Publications India Pvt Ltd, typeset in 10.5/12.5 pt Minion by Zaza Eunice, Hosur, India and printed at Chaman Enterprises, New Delhi.

Library of Congress Cataloging-in-Publication Data Available

ISBN: 978-93-515-0660-7 (HB)

The SAGE Team: Shambhu Sahu, Alekha Chandra Jena and Ritu Chopra

To
My wife Rashmi for her love and for persistently urging me on to write this book, and my sons Tanmay and Tatwam who are keeping the family tradition of loving books.

Contents

List of Tables   ix
List of Boxes   xi
List of Abbreviations   xiii
Preface   xix

1. Introduction   1
2. Trade of the Components of Biodiversity: Status and
   Challenges   11
3. Traditional Knowledge in the New IPR Regime   42
4. Implications of WTO and the North–South Divide   73
5. The Access and Benefit Sharing Regulations:
   Global Scenario   108
6. Case Studies on the Commercial Use of Biodiversity:
   Analysis of the Benefit Sharing Arrangements   141
7. Forging Bioprospecting Partnership Contracts:
   Mechanism for Equitable Benefit Sharing   206
8. Bargaining Power of the Resource Provider Countries   249
9. Resolving the Issues: The Way Ahead   263

References   297
Index   315
About the Author   321

List of Tables

5.1   Status of the Applications Received by
       the National Biodiversity Authority of India   124
5.2   Responses to Major Policy Concerns in ABS   128
5.3   A Comparison of the Provisions Made under
       the Agreement on TRIPS, the CBD and
       India's Biological Diversity Act of 2002   132

6.1   Case Studies Submitted for the Fourth
       Meeting of the COP to the CBD, May 1998   144

7.1   Sample 'Who Counts Matrix'   221
7.2   Influence Mapping of Various Stakeholders   222

List of Boxes

3.1 The Biodiversity Laws of Peru 48
3.2 Research and Information System (RIS) for
Developing Countries 55

4.1 The Agreement on TRIPS: Article 7 (Objectives) and
Article 8 (Principles) 83
4.2 The Agreement on TRIPS: Article 27 (Patentable Subject
Matter) 96
4.3 The Agreement on TRIPS: Article 29 (Conditions on
Patent Applicants) 100

6.1 The Aguaruna People of Peru and the Maya People of
Mexico 168

9.1 Cartagena Protocol on Biosafety 264
9.2 Bonn Guidelines 264
9.3 Doha Ministerial Declaration 2001 (Paragraph 19) 281
9.4 Nagoya Protocol 282
9.5 Nagoya Protocol on Access and Benefit Sharing:
Article 11 (Trans-boundary Cooperation) 295

List of Abbreviations

ABS	Access and Benefit Sharing
AICEP	All India Coordinated Ethnobotany Project
AVP	Arya Vaidya Pharmacy
BDCP	Bioresources Development and Conservation Programme
BGVS	Bedrijf Geneesmiddelen Voorziening Suriname
BHS	Biodiversity Heritage Site (India)
BIO	Biotechnology Industry Organization (Washington)
BMC	Biodiversity Management Committee (India)
B-MS	Bristol-Myers Squibb
BTG	British Technology Group
CAM	Complementary Alternative Medicine
CBD	Convention on Biological Diversity
CGEN	Conselho de Gestao do Patrimonio Genetico or Genetic Heritage Management Council (Brazil)
CGIAR	Consultative Group on International Agricultural Research (Centres)
CI	Conservation International
CIEL	Center for International Environmental Law
CIFOR	Center for International Forestry Research
CIPR	Commission on Intellectual Property Rights
CIR	Community Intellectual Rights
CIS	Commonwealth of Independent States
CITES	Convention on International Trade in Endangered Species of Wild Fauna and Flora
CNS	Central Nervous System

COP	Conference of the Parties (to the CBD)
CPs	Contracting Parties (to the CBD)
CSIR	Council of Scientific and Industrial Research (India/South Africa)
CSR	Corporate Social Responsibility
DFID	Department for International Development (UK)
ECOSUR	El Colegio de la Frontera Sur (Mexico)
EPO	European Patent Office
EPW	Economic and Political Weekly (India)
ETC	Erosion, Technology and Concentration
EU	European Union
FAO	Food and Agriculture Organization
FIC	Fogarty International Center (USA)
FIRD-TM	Fund for Integrated Rural Development and Traditional Medicine (Nigeria)
FPC	Forest Protection Committee
FPF	Forest People's Fund
FSI	Forest Survey of India
GATT	General Agreement on Tariffs and Trade
GDP	Gross Domestic Product
GOI	Government of India
GRAIN	Genetic Resources Action International
GTZ	German Corporation for International Cooperation
HFC	Healing Forest Conservancy
ICBG	International Cooperative Biodiversity Groups
ICFRE	Indian Council of Forestry Research and Education
ICIMOD	International Centre for Integrated Mountain Development (Kathmandu)
IDRC	International Development Research Centre (Canada)
IGC	Intergovernmental Committee (of WIPO) on Intellectual Property and Genetic Resources, Traditional Knowledge and Folklore
IIED	International Institute for Environment and Development
INBio	Instituto Nacional de Biodiversidad or National Institute of Biodiversity (Costa Rica)
INSTAR	International Network for Sustainable Technologies, Applications and Registration

IP	Intellectual Property
IPR	Intellectual Property Rights
ISMH	Indian Systems of Medicine and Homeopathy
ITPGR	International Treaty on Plant Genetic Resources
IUCN	International Union for Conservation of Nature
JNTBGRI	Jawaharlal Nehru Tropical Botanic Garden and Research Institute (India)
KIRTADS	Kerala Institute for Research, Training and Development Studies of Scheduled Castes and Scheduled Tribes (India)
KKSK	Kerala Kani Samudaya Kshema (Trust)
LDL	Low-Density Lipoprotein
LS	Laboratoires Serobiologiques (France-based multinational supplier of cosmetic actives)
MAP	Medicinal and Aromatic Plant
MAT	Mutually Agreed Terms
MBG	Missouri Botanical Garden
MFP	Minor Forest Product
MINAE	Ministry of Environment and Energy (Costa Rica)
MIRENEM	Ministry of Natural Resources, Minerals and Energy (Costa Rica)
MOU	Memorandum of Understanding
MTA	Material Transfer Agreement
NALSAR	National Academy of Legal Studies and Research
NAPRALERT	Natural Products Alert Database (Chicago)
NBA	National Biodiversity Authority (India)
NBF	National Biodiversity Fund (India)
NCI	National Cancer Institute (USA)
NFP	National Forest Programmes (of FAO)
NGO	Nongovernmental Organization
NHLBI	National Heart, Lung, and Blood Institute
NIAID	National Institute of Allergy and Infectious Diseases (USA)
NIC	National Informatics Centre (India)
NIDA	National Institute on Drug Abuse
NIH	National Institutes of Health (USA)
NIMH	National Institute of Mental Health (USA)
NNBP	Namibia National Biodiversity Programme
NSF	National Science Foundation (USA)

NWFP	Non-Wood Forest Product
OECD	Organisation for Economic Co-operation and Development
PBR	People's Biodiversity Register
PIC	Prior Informed Consent
PLA	Participatory Learning and Action
PRA	Participatory Rural Appraisal
R&D	Research and Development
RAFI	Rural Advancement Foundation International
RIS	Research and Information System for Developing Countries (India)
RRA	Rapid Rural Appraisal
SASI	South African San Institute
SBB	State Biodiversity Board (India)
SCBD	Secretariat of the Convention on Biological Diversity
SRISTI	Society for Research and Initiatives for Sustainable Technologies and Institutions
SUNS	South North Development Monitor
SWOT	Strengths, Weaknesses, Opportunities and Threats
TED	Trade and Environment Database
TIFAC	Technology Information, Forecasting and Assessment Council (GOI)
TK	Traditional Knowledge
TKDL	Traditional Knowledge Digital Library (India)
TKRC	Traditional Knowledge Resource Classification
TM	Traditional Medicine
TOI	Times of India
TRIPS	Trade-Related Aspects of Intellectual Property Rights
TWN	Third World Network
UK	United Kingdom
UN	United Nations
UNCED	UN Conference on Environment and Development
UNCTAD	UN Conference on Trade and Development
UNDP	United Nations Development Programme
UNEP	United Nations Environment Programme
UNESCO	United Nations Educational, Scientific and Cultural Organization

UPOV	International Union for the Protection of New Varieties of Plants
US	United States
USA	United States of America
USAID	US Agency for International Development
USDA	US Department of Agriculture
USPTO	US Patent and Trademark Office
VPISU	Virginia Polytechnic Institute and State University (USA)
WHO	World Health Organization
WIMSA	Working Group of Indigenous Minorities in Southern Africa
WIPO	World Intellectual Property Organization
WSSD	World Summit on Sustainable Development
WTO	World Trade Organization
WWF	World Wide Fund for Nature

Preface

The subject matter of this book caught my attention in the late 1990s when India's Biological Diversity Act was being drafted and every forester and scientist in India was talking about biodiversity. I was overwhelmed by the campaigns of the nongovernmental organizations (NGOs) on the issue of biodiversity and the ambiguities that surrounded it when my friend Padam Parkash Bhojvaid suggested me to take up this subject to study. My interest grew further when I had an interaction with Professor Samar K. Datta from the Indian Institute of Management, Ahmedabad, during a workshop in Shimla in 1999. Interacting with foresters and scientists frequently in seminars and workshops, I entered deeper and deeper into this field, making notes and shaping ideas, and Professor Datta encouraged me to articulate my ideas and shape them into a book form.

After about eight years of desk work, while nearing completion of this book, a question arose in my mind: What is the need of this book when thousands of pages have already been written on the subject? My subconscious mind replied: You have maintained the theme of the access and benefit sharing (ABS) as the 'central idea' and have deliberated to focus on the issues involved straightforwardly, posing the commercial use of biodiversity as a 'doable' action. The basic idea behind this book was generated from a few catchy words: 'Thinking globally, acting locally' (enunciated by Professor Charles R. McManis) and 'Given the high *hit-rate* in formal research around locally identified uses of biodiversity, costs of R&D in private and public systems are reduced considerably' (words of Professor Anil K. Gupta).

Additionally, various articles and reports written by authors who visited the sites where benefit sharing cases were taking shape helped me to bring out something which may work as a 'toolkit' for the natural resource managers involved in the business of biodiversity. In many of the forest-rich states of India, forest dwellers have been able to supplement their livelihood earnings by means of collection and marketing of the products of biodiversity with active intervention of government undertakings. It is woeful that the efforts made in the conservation and trade of local biodiversity through the joint forest management could not bring any big change. At this juncture, the 'bioprospecting' should be taken up. Now this is not happening because any 'bioprospecting partnership programme' would involve many agencies, require many leadership elements to be inculcated among the natural resource managers and demand hard work. Of course, large funding, tough negotiations and long years of project duration are some of the other prerequisites.

These prerequisites need not be looked upon as deterrents. The fact that a single word in the Agreement on Trade-Related Aspects of Intellectual Property Rights (TRIPS) could not be changed, even after more than a decade of continual international negotiations, also should not be seen as a deterrent. The road ahead is really tough and still there are many alternatives possible to take smaller steps to make a beginning.

Probably after reading this book, natural resource managers may find the mechanism of benefit sharing and the analyses given in this book of some use. Right here I intend to pose another question: Why should the natural resource managers of the tropical countries not turn into ABS professionals? The ABS should be developed for them as a new scientific-cum-legal discipline, incorporating the study of international conventions, biodiversity, biotechnology, marketing, local traditions and local laws, socio-economics, participatory rural appraisals (PRAs) and managerial techniques.

During the long years of writing this book, I have had the benefit of suggestions of Professor Samar K. Datta and Dr Padam Parkash Bhojvaid. I am also grateful to Professor Charles R. McManis and Professor Anil K. Gupta who supplied their references so quickly and whose analyses were most empowering. I am indebted to the international institutions—Center for International Forestry Research

(CIFOR), Food and Agriculture Organization (FAO), Secretariat of the Convention on Biological Diversity (SCBD), World Health Organization (WHO), World Intellectual Property Organization (WIPO) and World Trade Organization (WTO)—for the wealth of information they put on their websites.

The suggestions and support of Shambhu Sahu, Commissioning Editor of SAGE Publications, were great and gratifying. And lastly, I wish at least one simple case of benefit sharing, establishing a minis-cule 'trust fund' that can sustain a small local community, material-izes and let there be a salute to the ABS professional—and there are reasons to have optimism—high and high.

Shivendu K. Srivastava

1

Introduction

The commercial use of biodiversity of the world is an old phenomenon, and pharmaceutical industries have widely used the genetic resources and the traditional knowledge (TK) associated with these resources. Penicillin, quinine, atropine, menthol, taxol, morphine, salicin, borneol, digitalin and so on—the 120-plus list—are all natural product based. It was quite a phenomenon therefore that the last two decades of the twentieth century witnessed a sudden increase in the world's interest in the rich biodiversity of the globe along with the overpowering endorsement for the intellectual property rights (IPRs) related to genetic resources. This phenomenon started to show great concern for the conservation and sustainable use of the world's biodiversity and matured into what is now known as the Convention on Biological Diversity (CBD), followed by the Agreement on TRIPS under the larger umbrella of the WTO.

Out of these two international treaties, the impact of the second one—the Agreement on TRIPS—in the two foremost fields of the international trade, namely, agriculture and pharmaceuticals, has set off an incessant saga of deliberations, debates, demonstrations and declamations in India as well as in many other developing countries. Simultaneously, many partnership programmes that involved commercial use of biodiversity and associated TK began across the world post-WTO, a few of these prior to the WTO and even prior to the CBD. An exemplifying partnership programme was established between INBio (Instituto Nacional de Biodiversidad or National

Institute of Biodiversity), a nonprofit, semi-public organization instituted by Costa Rica's Ministry of Environment and Energy (MINAE), and Merck & Co., a US pharmaceutical corporation. With the prerequisite of an advance payment of the cost of genetic resources provided by INBio, the collaboration was viewed as the most lucrative deal because prior to this, Costa Rica earned nothing from its biodiversity. INBio worked in partnership with independent research institutions, universities and pharmaceutical, biotechnological, agro-industrial and cosmetic industries. The various agreements entered between them had arrangements for mutual benefits and had also assigned economic value to Costa Rica's natural resources.

The example of INBio, which entered into dozens of other collaborative programmes and became a leader in the negotiation and benefit sharing agreements, with thousands of pages written about, is not the only case. Today, there exist many partnership programmes having multilateral agreements, and a significant one is the International Cooperative Biodiversity Groups (ICBG) programme, which premises on the belief that the discovery and development of pharmaceutical and other useful agents from natural products can promote scientific capacity development and economic incentives to conserve the biological resources from which the products are derived. Launched in 1992, the ICBG programme is jointly funded by the US National Institutes of Health (NIH), National Science Foundation (NSF), National Oceanic and Atmospheric Administration, US Department of Agriculture (USDA) and Department of Energy. Under the parasol of the ICBG programme, several projects were taken up across continents. Each project was tailored to the particular conditions of the countries involved and addressed three issues: drug discovery, biodiversity conservation and sustainable economic growth.

Another case of commercial use of biodiversity and sharing of the benefits accruing from this use with the indigenous people was made to happen by the Bioresources Development and Conservation Programme (BDCP, an NGO working in Nigeria), which facilitated the establishment of the Fund for Integrated Rural Development and Traditional Medicine (FIRD-TM) in Nigeria. The institution of FIRD-TM has a goal to build technical skills, using the biological resources as a means for improved health care and sustainable development. It receives funds from many contributors and channels the benefits in an equitable and consistent manner to the resource provider communities from which commercially useful biological resources and ethnobotanical knowledge are derived. It was registered under the relevant local laws so as to give it a legal entity capable of owning property and maintaining and defending its actions. It is also unique in the sense that its research programme also targets therapeutic categories for tropical diseases such as malaria, leishmaniasis and trypanosomiasis suffered in Nigeria. The mechanism of trust fund also, of this archetypal case, is unique in the sense that it lays down an excellent way for sharing of the benefits accruing from the commercial use of biodiversity.

These and other few creditable cases of ABS, discussed in detail in Chapter 6, have complex multilateral agreements. The IPR issues related to the associated TK were treated under these cases in different ways—varying from joint ownership of patents and joint authorship of publications to exclusive right to draw patents. Sharing of the royalty and the cost of patenting and patent protection also constituted a part of these agreements. Compared to these, the case of sharing of royalty earned from making use of TK by the Jawaharlal Nehru Tropical Botanic Garden and Research Institute (JNTBGRI)[1] in Kerala (India) constituted a very simple agreement. When JNTBGRI developed the energy-giving drug *Jeevani* from Arogyapacha (*Trichopus zeylanicus*), a local perennial herb, making use of the TK provided by the indigenous Kani people, and obtained a patent for the drug, it stipulated to share 50 per cent of the fee and royalty from the drug with the Kani people through a trust fund constituted for local developmental initiatives and biodiversity activities. The work of the trust has led to employment and income generation activities. Nevertheless, the benefit sharing arrangements in this example have ridden a rough road because of poor coordination between various stakeholders. It is learnt that the case of benefit sharing has not died and efforts are on at the JNTBGRI to involve an institution of the Kerala Forest Department in order to steer a fresh set of agreements that would involve and benefit all the major stakeholders.

There is a new case in point for the purpose of benefit sharing that adopts a corporate social responsibility (CSR) approach. The unique case from Morocco (the country does not have any ABS legislation)

---

[1] Previously known as TBGRI; visit: www.jntbgri.in.

involves the supply of argan (*Argania spinosa*) oil and oil cakes with some other products through local women's cooperatives to the international cosmetic industry. The research component engaging a number of national and international partners has the objective of investigating the characteristics of the oil and other derivatives of argan, predominantly to develop cosmetic products. The benefits are paid through a social fund, which the women's cooperatives may well use for whatsoever diverse purposes they wish.

The accrual of benefits from the argan oil of Morocco, from Arogyapacha of India, from the wide range of extracts from Costa Rica or the benefits gathered in the form of drug discovery, biodiversity conservation and sustainable economic growth being brought in through programmes such as ICBG or BDCP undertaken in many tropical countries raises ample hope and enormous prospects. The expectations are expressed at various platforms by developing countries, and the stronger the voices, the more caustic become the debates on the issue. The debates often turn into serious conflicts between the developed countries (with poor biodiversity but high technology) and the developing countries (technologically poor but rich in biodiversity). The following comment from developed country professors, Simpson and Sedjo (2004), could prove how bitter the conflict has become:

> The biggest practical problem with bioprospecting may be unrealistic expectations. *Our sense* is that governments and advocates in developing countries have often been *misled* about the value of their biological wealth. Consequently, they hold out for *spectacular gains that they are unlikely to achieve.* (Emphasis added)

This conflict originates from the key provision of Article 27 under the Agreement on TRIPS that articulates the favour for granting patents to the technologically advanced developed countries, whereas it puts the rich biodiversity of the tropical countries at the risk of biopiracy (see Chapter 3). On the other hand, the CBD has provisions for equitable sharing of results of R&D and the benefits arising from the commercial use of biodiversity. In between the two—the Agreement on TRIPS and the CBD—many countries are in the process of framing ABS legislations suited to their specific needs and domestic laws. Concurrently, the partnership programmes that began with the

INBio–Merck collaboration the world over and greatly involved the commercial use of biological resources of the source countries put up textbook-type examples of action that may very well be expressed as 'thinking globally, acting locally' if we use the phrase put forth by Charles R. McManis (2003) of Washington University, St Louis (USA).

It is found that one marked approach in general in these 'thinking globally, acting locally' cases of partnership programmes involving bioprospecting has been to provide economic incentives to the indigenous and local communities as well as skill development and infrastructural support for collection and processing of the resources. The indigenous and local communities in all partnership programmes are recognized as primary or on-site stakeholders, and the TK they own is extensively exploited in most of the commercial uses of biodiversity. It is principally affirmed that they must have full usufructs over adequate natural resources, located at an accessible distance, so that their basic needs are fulfilled (Singh et al., 2000).

The idea of providing the usufructs, though it raises ample hopes and enormous prospects in the light of successful experiences of benefit sharing, is difficult to implement in the real sense, yet the opportunities look significantly bright. The domestic herbal industry in India, for example, is quite strong, giving rise to reasonable expectations, and the TK system developed through generations of traditional healers has vast scope in bioprospecting that frequently uses advanced biotechnological processes and techniques owned in general by the developed countries. Unfortunately, the countless efforts aiming the distribution of usufructs and benefit sharing exhibit only a top dressing approach, and the core benefit from the mine of biodiversity prospecting has yet to accrue to the indigenous and local communities. Another point of concern, as Professor Anil K. Gupta (2005) of the Indian Institute of Management, Ahmedabad, cites, is that the available incentives and the community structures have failed to motivate young people to carry the torch of knowledge forward for the next generation, and the erosion of knowledge and resources has never been as high as it has been in the current generation.

Howsoever emphatic the pronouncements underscored in the preceding paragraphs may be, the reality is that the Agreement on TRIPS neither recognizes the contribution of TK nor contains any provision allowing any claim to enforce fair and equitable sharing of benefits that accrue from the commercial use of the genetic resources.

When the stipulations of the Agreement on TRIPS are discussed vis-à-vis the sharing of commercial benefits with the industries and the corporations, especially the clause of patentable subject matter, it is argued that the corporations not only take away the resource that the indigenous communities have the first right to, but also multiply their profits. This happens because the application of the TK associated with these resources and held by these communities leads to a higher hit-rate[2] in screening of the active ingredients as compared to the random sampling method. The other contention, based on the verity of socio-economic dependence of the indigenous people on these resources, is that unless an alternative for subsistence and income is provided to these people, it would be inappropriate to deprive them of the share in the benefits accruing from the very resource and the use of the associated TK held principally by these people.

In continuation, the issue of how the interests of the biodiversity-rich developing countries must be safeguarded needs to be examined along with the most fundamental question as to why at all the benefits should be shared with the indigenous and local communities. It is certain now that the resoluteness of the provision under Article 27 of the Agreement on TRIPS will depend upon the unity of purpose and the bargaining power of the tropical countries that they are able to draw from the CBD. The biological diversity that resides on their land masses, instead of being treated as common property resource of the indigenous communities, is recognized now as the common heritage of the whole of humankind, and this treaty ought to be perceived in a much broader perspective if an equitable sharing of the benefits of commercial use of biodiversity is set as the goal.

In the context of the ownership issue of natural resources, during the recent years, one distinct change has come up—the emphasis now is on the state to have the ownership of the rights over the biological resources, and the state only as promulgated by the CBD has the authority to determine access to these resources, including the authority to make necessary laws to this effect. By and large, though there is stress on biodiversity conservation, including the protection of TK, innovations and practices of indigenous and local communities

---

[2] It refers to that percentage of samples that show biological activity in a given series of chemical screens.

and equitable sharing of benefits accruing from these resources, the people (the indigenous and local communities) in general remain sidelined unless, of course, they are involved as the primary stakeholders in the process of partnership programmes. In more explicit terms, it amounts to say that the right of ownership of the genetic resources, which ought to rest in the people, rests in the state, and it is now left to the wisdom of the state how far it allows this right to be exercised in practice to an extent that the objectives of the equitable benefit sharing, as mandated under the CBD, are met.

If in the above background, the developing countries adopt a common approach in negotiating with the developed world, an acceptable equilibrium may develop in such relationships despite the language in the CBD not being mandatory on these issues. Article 27.2 of the Agreement on TRIPS recognizes that the states can exclude from patentability inventions, the prevention of whose commercial exploitation is necessary to protect human, animal or plant life or health or to avoid serious prejudice to the environment. It follows that an invention, the commercial exploitation of which is permitted by domestic laws, cannot be excluded from patentability.

In other words, if commercial exploitation of the natural resources of a country is allowed under any arrangement, then the invention related to that cannot be excluded from patentability in any part of the world. Doha Declaration on the TRIPS Agreement and Public Health (WHO, 2001b) also talks about the gravity of the public health problems of developing countries and yet it is silent about the protection of TK. The Declaration recognizes that in order to promote access to medicines for all, each member country has the right to grant compulsory licence and freedom to determine the ground upon which such licences are granted, though the aspect of TK is not alluded to. It is only the Nagoya Protocol on Access to Genetic Resources and the Fair and Equitable Sharing of Benefits, adopted by the Conference of the Parties (COP) to the CBD at its 10th meeting in 2010, that provides a framework for implementation of the ABS provisions set out in the CBD and paves the way ahead for sharing of the benefits from the utilization of genetic resources in a fair and equitable way, including by appropriate access to genetic resources and by appropriate transfer of relevant technologies. The Nagoya Protocol, which entered into force in October 2014, may provide a transparent legal framework for the effective implementation of the third objective of the CBD (fair and equitable sharing of the benefits arising out of the utilization of genetic resources). The Protocol has three main pillars—A, B and C—described below[3]:

- **Access**: Establish more predictable conditions for access to genetic resources.
- **Benefit sharing**: Ensure benefit sharing between users and providers of genetic resources.
- **Compliance**: Ensure that only legally acquired genetic resources and TK are used, complying with domestic legislation.

The ongoing deliberations on these aspects generate a lot of optimism on the part of the developing countries, though often the issue of protection of TK is still eluded. Meanwhile, the tropical countries are in the process of evolving sui generis system to undertake measures to protect the knowledge of local people relating to biodiversity that may include the issues of access to the biological resources, registration of the TK as well as the benefit sharing arrangements at local or national levels and other measures for protection and sustainable use of their biological resources.

This scenario depicts nothing but what is now termed as globalization, and comprehensively, the arrival of the WTO is an obvious pronouncement of the same. This broad framework, together with the provision under Article 41 (*IPR enforcement not to be unnecessarily complicated or costly and not to entail unwarranted delays*) of the Agreement on TRIPS, may constrict the domestic laws related to the protection of TK being legislated by the tropical countries. That is the challenge before the tropical countries, and the countries that seemed to act in unison recently (viz. Brazil, South Africa and India) could take a lead to bravely confront this situation.

Taking up the challenge intently, the tropical countries are in the process of formulating such laws. Many have already enacted the laws seemingly capable to protect their genetic resources as well as to ensure equitable sharing of benefits. The enforcement of such laws has

---

[3] Heard first about the 'ABC' of Nagoya Protocol from Dr Archna Negi, School of International Studies, Jawaharlal Nehru University, New Delhi, in a workshop held at Indian Society of International Law, New Delhi (22–23 August 2013).

yet to be made effective in several other tropical countries, and, the terms such as protection of TK or declaration of the place of origin of a genetic resource utilized in the new research works remaining unspoken in the Agreement on TRIPS, these countries shall have to tread a rough road ahead if they aim an equitable sharing of benefits accruing from the use of their biodiversity. Furthermore, their biodiversity laws should not attract the blame of putting distortions and impediments to international trade or laying down procedures that become barriers to trade. The Agreement on TRIPS by itself basically decrees to reduce distortions and impediments to international trade and to ensure that the procedures to enforce IPRs do not themselves become barriers to trade.

This aspect essentially complicates the underlying issue, as any country that endorses the WTO cannot choose the Agreement on TRIPS to be unendorsed. The only support at this juncture comes in view from the unambiguous though soft-worded mandate of the CBD (*Article 8j*: Knowledge, innovations and practices of indigenous and local communities to be respected, preserved and maintained, and promoted for wider application; each contracting party [CP] to encourage equitable sharing of benefits arising from utilization of such knowledge, innovations and practices).

The ambiguities and the conflicts be there, the interest continues even today in the wealth underneath the rich biodiversity of the globe, and against this backdrop are witnessed many international contracts and collaborations involving biodiversity-rich communities and regions, technology-rich nations, pharmaceutical industries and corporations, universities and research institutions to take up the commercial use of biodiversity of the world in a big way. These contracts largely follow the central ideology of how to share the benefits from exploitation of the genetic resources and TK associated with these resources.

Spontaneously, the trend is to circumvent the IPR regime and collectively work on a mechanism that merely ensures equitable sharing of benefits accruing from these resources. So far as the benefits are shared in reality, the patentability issues may be sidelined as what is ultimately aimed at in the IPR regime also is the sharing of benefits only. Simultaneously, methodologies are being evolved to draw the framework of an equitable benefit sharing mechanism. Incidentally, a few recent happenings indicate, though not clearly, a decline in the level of interest of large pharmaceutical corporations in bioprospecting, and the reason attributed to is the complexities in the benefit sharing mechanism. The withdrawal syndrome on the part of these firms could be attributed to the attitudinal protectionism that is common in manufacturing industries. In the above background, it is very unlikely that a methodology is evolved to draw a 'one size fits all' mechanism for benefit sharing. Also, the bioprospecting partnerships being worked out are highly technical, complex and legal in nature, and they involve numerous institutions, tough negotiations and they work under several mega-influences. Therefore, every bioprospecting programme needs to be treated as a different case for drawing the benefit sharing mechanism.

A need arises certainly for symbiosis between the Agreements on TRIPS and the CBD. Conformity with the CBD could be supported by a liberal interpretation of the Agreement on TRIPS, especially Articles 7 and 8, which surely does not seem to be possible in reality very soon. In the mean time, the biodiversity-rich and resource provider tropical countries are looking for the tool that could safeguard their rights, and new Biodiversity Acts and ABS regulations are being formulated following continual debates that include diverse perceptions of different stakeholders. The forthcoming chapters in this book analyze the examples across the world and examine their complexities to offer workable options for ABS in the future.

The analysis of the case studies and the international IPR regime, followed by an attempt made in this book to find out how to circumvent the implications arising due to the present limitations, should come in handy as a good working tool to take up new bioprospecting programmes in the tropical countries. The task is 'easily said than done', and yet, if following the 'thinking globally, acting locally' concept, the stakeholders sit together and make sincere efforts, more replications of the benefit sharing cases can be formulated and enacted. Let us be optimistic, read on and take the lead.

2

Trade of the Components of Biodiversity: Status and Challenges

Since prehistoric times, the natural resources of the forests in developing countries have provided people with a rich supply of food and other basic needs of life. The products that people gathered for subsistence comprised wood as well as non-wood products, a differentiation that is more significant to foresters than to forest dwellers. Prior to 1990s, the non-wood forest products (NWFPs) were forthrightly called as 'minor forest products' or MFPs, while timber was supposed to be the major revenue-earning product from the forest. After protracted debate that these products should be termed as non-timber or non-wood products, replacing the term 'MFPs' with NWFPs, the jargon was settled in 1991 when the FAO established the 'Non-Wood Forest Products Division' under the Forestry Department of FAO.

The NWFPs, forming the main group of genetic resources that are extracted from natural forests, include medicinal and aromatic plants (MAPs) too. The forest dwellers gather these products from the natural forests for their consumption and also to sell in the local weekly market to the local traders or middlemen, bartering in general for cash or for common items of domestic consumption. The role and contribution of NWFPs are crucial among forest dwellers and other village communities, particularly in developing countries, because of their richness of variety as a source of sustenance and livelihoods as well as the storehouse of traditional health-care systems. As reported, some 80 per cent of the populations of developing countries depend on NWFPs for their primary health, nutritional needs and income generation (FAO, 1995; Farnsworth et al., 1985; Maudgal and Kakkar, 1996).

An estimation of the WHO also states that about 80 per cent of the world's population is dependent on medicinal plants for primary health care, and this figure remains to be quoted frequently with sanctity (Meena et al., 2009; Prashantkumar and Vidyasagar, 2008; Verma and Singh, 2008), while some other figures also, though supportive to it, are available. Martins Ekor (2013) states that up to four billion people representing 80 per cent of the world's population living in the developing world rely on herbal medicinal products as a primary source of health care. Wachtel-Galor and Benzie (2011) state that in Africa up to 90 per cent and in India up to 70 per cent of the populations depend on traditional medicine (TM) to help meet their health-care needs. According to Kamboj (2000), herbal medicine is still the mainstay of about 75–80 per cent of the world's population, mainly in the developing countries, for primary health care.

The statistics quoted above are an indication as to how significant the herbal medicines are to the world's populations. Lately, new acronyms have been assigned to the group of herbal medicinal products that are usually referred to as TMs. TM now stands for traditional medicine and CAM stands for complementary alternative medicine, together termed as TM/CAM. The term CAM is used outside India, and WHO refers to it as TM (Mehrotra, 2003).

The components of MAPs, in addition to serving the needs of primary health care and nutrition, also help in the earning of wages through the collection and sale to the local traders. Simultaneously, numerous other NWFPs provide the raw material for varieties of artefacts and other utility items that the members of forest dweller families make in order to earn cash income, though little. Collection of these products from the wild is a source of cash income especially during the lean season when they find lesser engagements in the farm work, and out of the total population of India, about 100 million people living in and around forests derive at least part of their livelihood from collection and marketing of NWFPs (Kumar et al., 2000; Saxena, 2003).

Consequently, the issue of rights and access to these resources and the income accrued from them is of great importance to these communities. These products are not confined to the natural forests only, as Iqbal (1995) mentions while describing what constitutes NWFPs:

> NWFPs are considered to be as all the biological materials (other than timber and firewood) that may be extracted from natural ecosystems, managed plantations and semi-wild trees growing on farmlands and be utilised within the household, be marketed, or have social, cultural or religious significance. Both plant and animal products are included.

While the above definition adequately covers each and every aspect one would require it to include, FAO gave a precise definition. Based on the recommendations of an internal interdepartmental FAO meeting on definitions of NWFPs held in June 1999, the following new FAO working definition of NWFPs was adopted (FAO, 1999):

> Non-wood forest products consist of goods of biological origin other than wood, derived from forests, other wooded land and trees outside forests.

If the above definition appears too concise to discuss the aspect of commercial use of biodiversity, the definition articulated by Iqbal is more suitable. Going further, let the constituents of NWFPs be ascertained. The NWFPs are used for their medicinal, cosmetic or cultural purposes and also as food and food additives such as edible nuts, mushrooms, fruits, herbs, spices, resins, gums, plant exudates, bark products, cork and so on and include several animal products. Though firewood, minerals and services such as wildlife tourism recreation and so on, which are attributed to the forest ecosystems, also may be taken as constituents of NWFPs, the aspect of equitable commercialization vis-à-vis international conventions can be discussed with reasonable justification within the scope of the definition of NWFPs as adopted by FAO.

Even if we emphasize the commercial value of ecosystem services, the way the biodiversity is of use presents two different aspects as regards its values. Let us keep the humankind in the centre of the biological diversity that surrounds it and forms the ecosystem in which the humankind stays alive. The two aspects that are far apart as regards the values of the biodiversity are the direct and the indirect values. The ecosystem services of the biodiversity and the ecological functions such as climate regulation, soil and water conservation and so on will constitute the indirect values of biodiversity as these values are shared naturally and quite fairly (of course ignoring the externalities of production economy). As against this, the edible products, MAPs or all the genetic resources or the NWFPs that form the resource base for basic sustenance and livelihood for forest dwellers hold the direct values of biodiversity. The unsustainable use or indiscriminate exploitation of the biological diversity today vis-à-vis both the direct and indirect values would affect the future generations, but surely the commercial use of the products of biodiversity that have direct value bear more significance particularly in the context of ABS issues.

The commercialization of these natural resources spans over a wide range—from village level to the international level—as the products gathered by the local people, apart from the household and healthcare uses at the village level, find use in various types of industries, for example, pharmaceutical, herbal medicine, personal care, cosmetics, horticulture, crop protection and biotechnology, as the end products from these industries enter international markets. At the global level, the total import of NWFPs is of the order of $11 billion in terms of its value, of which about 60 per cent is imported by developed countries: USA, Japan and the European Community (Iqbal, 1995).

Apparently, particularly in the context of international trade and the new IPR regime, the term 'NWFP' has begun losing its connotation now. The more commonly used and probably more relevant terms are 'biodiversity', 'natural resources', 'biological resources' or 'bioresources', and quite often 'genetic resources' too. Strangely enough, the import of these resources creates work opportunities even in the developed countries where new uses of these products are being found every day particularly with the rising preferences for products drawn from a plant base compared to the chemically synthesized products. In addition, gathering of high-value products such as mushrooms (morels, matsutake, truffles) and medicinal plants (ginseng, black cohosh, goldenseal) also continues in developed countries for cultural and economic reasons (Jones et al., 2002). Nevertheless, collection, processing and local trade of these resources are major occupational activities for the rural people of developing countries only. The manufacturing, trade and marketing of these resources certainly involve the people from both the developing and developed countries.

Within the national boundaries in India, the trade of genetic resources is highly secretive, uncontrolled and unorganized. The international trade, however, is controlled and regulated by various trade measures. The trade is regulated by both the tariff-related and the non-tariff-related measures. The developed countries have more tariff regulations over export of the finished goods, while more liberal is the policy of import of natural resources coming mostly from the 'South' (Iqbal, 1995). The developing countries have such provisions the other way round. Interestingly, new market preference in general for natural products has enhanced the demand of genetic resources in national and international trade. Overall globalization in the trade of pharmaceutical and agricultural produce, removal of the quantitative restrictions on import and opening of national economies have much wider impact on the commercial use of these resources.

Investigations into the commercial uses of the components of biodiversity at the global level make it evident that the general flow of international trade in these products is directed in general from the biodiversity-rich and technology-poor southern regions of the continents or the 'South' to the biodiversity-poor and technology-rich northern regions of the continents or the 'North' (Iqbal, 1993; Khoshoo, 1995; Vantomme, 2001; Walter, 2002). As regards the national-level trade-related data, the Customs Department in India categorizes plant exports as crude drugs or bulk drugs and, more often than not, does not record species. The expertise available at export points for identification of plant products and derivatives is also very minimal (Menon, 1996). The lack of expertise and the related unawareness on these issues might prove dear to the tropical ecosystems. Before examining the complexity of this scenario vis-à-vis the regulations of WTO as well as the new IPR regime, it would be pertinent to make an assessment of the wealth of the genetic resources and an analysis of the international trade of their various components.

Status of the Biological Diversity

As has been expressed above, broadly speaking, the terms 'biological diversity or biodiversity', 'bioresources', 'genetic resources', 'natural resources' or 'NWFPs' refer basically to the resource base wherefrom the rural people gather various products for their health-care needs, sustenance and livelihood. Though the products are derived from forests, other wooded land and trees outside forests, the bulk of the collection in general is from the wild (Kuipers, 1997; Maudgal and Kakkar, 1996) and, passing through an intricate channel, ends up in manufacturing industries in the form of either dried parts of plants or plant extracts as raw material. For example, the Indian drug and pharmaceutical industry continues to get almost 90 per cent of its supplies from the collections made from the wild (Sarin, 2003b). The dependence of the manufacturing industries on the collection from the wild is because of the hard facts of the economics involved. In an FAO technical paper, Kuipers (1997) records that there are several reasons attributed to this phenomenon, notably the low cost of wild harvested material, and the fact that cultivated material requires management and agricultural expertise, time (sometimes more than 10 years before the crop is ready for harvesting), land, financial resources and so on before an income can be derived. Another characteristic is that the MAP-based industries significantly depend on these resources and draw heavily from the field of knowledge related to these products that is held by the indigenous people living inside or on the fringes of the forests. The statement that out of the 119 plant-derived drugs listed by Farnsworth et al. (1985), 74 per cent were discovered as a result of chemical studies to isolate the active substances responsible for the use of the original plants in TM proves that scientists have not hesitated in drawing upon the abundant and culturally acceptable local knowledge when it was appropriate. This aspect has evoked immense interest among them to take up scientific and industrial research in the field of knowledge related to these products.

The field of knowledge in common parlance is referred as TK, which shall be discussed later. Firstly, an assessment of the status of the biological diversity is attempted in this section with special reference to the developing countries. As proven frequently, the tribal or the aborigine populations geographically overlap with the forest areas in a country. As per the India State of Forest Report 2013 (FSI, 2013), 189 districts spread over 26 states and union territories in India, identified as tribal districts by the Government of India (GOI) under the Integrated Tribal Development Programme, have total forest cover of 415,491 sq km, and although these districts constitute only 33.82 per cent of the total geographical area of the country, the forest cover in these districts constitutes 59.53 per cent of the total forest cover of India. Put simply, about 60 per cent of the forest cover of the country is confined to only about one-third of the geographical area of the land that is inhabited by tribal communities.

This plainly indicates that the tribal populations are concentrated in the locations rich in forest resources. The forests have traditionally played a vital role in the economy of tribal people, yet the forest dwellers are the laggards in the socio-economic development process, living in primitive conditions. This is corroborated well by I. Koziell and Charles McNeill (2004) who find that many areas bearing the world's most valuable biodiversity coincide with populations of abject income poverty and social marginalization, where poor people are overexploiting local resources, forced to subsist from areas or resources too small and too unproductive to properly support a sustainable existence.

These observations exhibit, on the one hand, a linkage between the socio-economic dependence of the forest dwellers with the status of biodiversity of the regions these people belong to and, on the other hand, the relevance of the very definition of NWFPs (that the biological materials have social, cultural or religious significance to the forest dwellers). Very logically, it is frequently stated that the indigenous communities are culturally dependent on forests for their subsistence and livelihood needs, and the trade of the products from the forests in turn is dependent on the indigenous communities as to how efficiently they are gathering the products from natural forests. They have an expertise as to where to find and how to identify a particular plant in the wild, which part of the plant to pick and choose, the right stage of collection of a particular part of plant, awareness about its habitat and often the basic extraction methodology for many species of the wild.

The socio-economic and cultural dependence of the forest dwellers on biological resources and the quantum of trade of the components of biodiversity have direct bearing on the status of biodiversity of the tropical countries. As regards the status of MAPs in India, documenting the wealth of the medicinal plants, Sarin (2003a) emphasizes that there has been a gross depletion of the natural population of a number of medicinal plants, quite a few of which have become vulnerable while at least 10 are endangered and on the verge of extinction. Sarin (2003a) has enlisted 146 species under the category of 'Medicinal plants growing in forests, grasslands, running or stationary water bodies, deserts and other forms of natural vegetation', 53 species in the category of 'Medicinal plants growing as weed or under run wild conditions in secondary forest scrub, fallow agricultural land, orchards, organic dumps, along rail track or roads, in and around stagnant water bodies and other waste places', 30 species under the category of 'Plants cultivated as avenue trees, embankment stabilizers, hedges or ornamentals in parks and gardens and yielding herbal raw materials', 40 species under the category of 'Plants grown as agricultural, horticultural or industrial crops and also yielding important herbal raw materials' and 29 species in the category of 'Plants cultivated exclusively as medicinal crop'. He has also enlisted 40 species that are imported entirely or partially as raw materials from other countries (except Nepal and Bhutan) to supplement indigenous production. He states that out of 30 medicinal plants under cultivation, at least 10 are being gradually abandoned or replaced by other crops that are more paying or have a regular market.

In the context of the status of biodiversity of other tropical countries, it is contended that the forests of South East Asia have traditionally remained the major source of many of the natural products. It is established that India, Indonesia, Malaysia, Thailand and Brazil are major suppliers of the raw material to the world market, with China reaching new heights to dominate the world's trade (Sharma et al., 2001). Incidentally, India rightfully attains prominence in the South and South East Asian region as regards the trade in genetic resources, as it is the fifth largest manufacturer of bulk drugs and also one of the 17 mega biodiversity countries[1] of the world. As a document submitted by India to the WTO mentions (WTO, 2000), with only 2.4 per cent of the world's land area, India accounts for 7–8 per cent of the recorded species of the world (based on the survey of 65–70 per cent of the total geographical area of the country). The document maintains that the Botanical Survey of India and the Zoological Survey of India have recorded over 47,000 species of plants and 81,000 species of animals, respectively, and it is anticipated that some of the remaining areas (e.g., Himalayan region and Andaman and Nicobar Islands) may be far richer in biological diversity than most of the areas already surveyed.

---

[1] *Source*: National Biodiversity Authority (NBA) of India; visit: www.nbaindia.org

While it is evident that the overall picture of the biodiversity that exists in the developing countries is not so dismal, when the trade aspect of the components of biodiversity is investigated, it is observed that the job is a difficult one because the data related to the trade of the components of biodiversity are too scanty, as we find in the following section.

Scanty Data of Trade in Biodiversity

As is so in other tropical countries, it is recorded that India is equally rich in traditional and indigenous knowledge, both coded and informal, but still the trade data are scanty. The Ernst & Young has valued the Indian pharmaceutical market at $4.5 billion, where the domestic market is growing at an annual rate of 8–9 per cent per annum, and more than 25 per cent of the finished medicines, valued at $3 billion, are exported to countries such as the USA as well as Russia and other Commonwealth of Independent States (CIS) countries.[2] Though such information about the magnitude of trade and the resource base are available, it is believed that the information that exists about the resource base, production, consumption and trade of the products of biodiversity is inadequate and scattered. The lack of accurate information is believed to be the biggest constraint experienced by researchers with regard to the study of the natural resources of the tropics, despite the widespread usefulness of the products and their value in the international market. It is regrettable that there are no reliable data even on the number of plant species that are currently traded in high volume, while such a list is badly needed (Kuipers, 1997). The FAO also validates that the availability of specific information on the status and trends in forest genetic resources is currently woefully inadequate (FAO, 2010). The Global Forest Resources Assessment 2010 of FAO cites that there is no consolidated global picture on the status and trends of forest genetic resources, and estimates of the rate of genetic diversity loss are lacking. It is heartening, however, that the State of the World's Forest Genetic Resources

---

[2] Indian pharma market worth $4.5 b: E&Y. *Economic Times*, New Delhi, 26 December 2002.

is envisaged to be prepared through a country-driven approach based on country reports and thematic studies.

In the same context, it is also remarked (Brown, 1995) that the network for trade in medicinal plants is informal, and that characterizing the trade is difficult. The statistics incorporates a large range of products and issues and no single database is available covering all products and all aspects. Tan et al. (1996) observe that some databases are more general and comprise many products, while others deal only with few specific products and span over a short period. Moreover, collection of primary data involves high costs, and quantitative data are harder to come by. Many researchers have examined as to what could be the reasons behind the lack of adequate data related to the natural resources. Chandrasekharan (1995) enumerates the following reasons for the near absence chiefly of NWFPs in official statistics or national accounts, wrongly indicating a relatively low contribution of forests to gross domestic product (GDP) and national welfare:

1. NWFPs form a heterogeneous group;
2. Transactions related to these products largely take place in households and small-scale units; and
3. While NWFPs are very important in local economies, they mostly form part of an informal sector and are outside the established marketing system.

The state forest departments in India also do not keep sound data as regards the quantities removed from the government forests, the sale value of these products or the status of the resource base. The products gathered locally by the indigenous communities are removed from the forests through the channel of local traders acting as middlemen, and if removed in small quantities without the use of a transport vehicle, and that is generally the trend, no permit is required for the transport of the produce, and it goes out unrecorded. Local traders and middlemen enjoy free access, at par with the indigenous communities, without paying any royalty. This leads to haphazard removal of the fruits of biodiversity without any record kept as regards the quantity removed. The motive behind providing free access to these resources is that the indigenous and local communities who gather these products in order to earn livelihood should get better prices for what they gather, and therefore no royalty is charged for the material collected from the forests. It is wondered whether these reasons are justifiable.

On further investigation of the issue, it is witnessed that the products of biodiversity are basic resources utilized by many high-valued industries such as pharmaceutical, herbal medicine, personal care, cosmetics, agrochemical, horticulture, crop production and biotechnology. The resources often originate from tropical regions, where they are found in greatest diversity, and are collected from lands inhabited by local communities and, making use of the TK, the potentially valuable resources are identified and commercialized. Down the line, the resources fetch high value, particularly when the final products are patentable subject matter. Though it is usually seen that the future drugs, industrial products and genes for improved crops derived from the natural resources are being sought predomi-nantly in the genetically rich developing world (WIPO, 2004b), the benefits accruing from these resources are not shared at all with the providers of the resource. Evidently, when the trade channels are kept secret and there is no sharing of benefits, no trade data are generated or recorded at all.

These observations indicate why the biodiversity trade data are scanty despite the high commercial value of the components of bio-diversity. Lesser (1998) observes that the technology, including bio-technology, enhances the commercial value of the components of biodiversity, and many mechanisms underlie this increase in eco-nomic value, including the option to transfer genes directly among unrelated species, the speeding up of techniques for screening materi-als for possible medicinal effects and a general increase in the rate of bringing materials to the marketplace. But whatever the mechanism, the significant point is the increased commercial value due to the use of genetic resources. This state of affairs has bewildered the stakehold-ers in developing countries, raising concerns for the conservation and the future availability of genetic resources as high commercial value of biodiversity may lead to unsystematic and destructive exploitation.

Manifestly, it is found that the data related to the trade of the components of biodiversity are scanty, and simultaneously a fear looms large that the valuable indigenous wealth will be taken away and exploited commercially by the transnational pharmaceutical companies (Patnaik, 1997), with little or no returns to the countries providing these resources. As per an estimate of the United Nations Development Programme (UNDP), medicinal plants and microbials from the 'South' contribute at least $3 billion a year to the 'North's pharmaceutical industry (Nair, 2000). Developing countries need to capitalize on their unique bio-assets if they have to become economically strong. They need to have systems that will offer benefits from the trade of the medicinal plant resources. These and other related issues in the context of international trade of the biological resources are discussed in the following section.

International Trade of the Components of Biodiversity

The wide range of the products of biological diversity extending from fruits and nuts to aroma chemicals and phytopharmaceuticals leads to their use in a wide extent of markets at the local, national and international levels, as well as for bartering in subsistence economy (FAO, 1995). As regards the international trade, as has been established in a previous section, the general flow of trade of these products is directed mostly from the biodiversity-rich and technology-poor 'South' to the biodiversity-poor and technology-rich 'North'. In this context, McManis (2004) perceives a growing international awareness of the link between the development of biotechnology and the preservation of genetic resources, as technology-rich industrialized countries, which are spearheading the development of biotechnology, are conversely discovering that they are relatively biodiversity-poor, while developing countries, although technology-poor, are beginning to realize that they are the stewards of the bulk of the earth's biodiversity. This North–South divide has demarcated the positions of the technology-rich developed world and the biodiversity-rich developing world. Iqbal, one of the earliest investigators of this divide, wrapped up his study by summarizing his results (Iqbal, 1993) that depict that main origin for the direction of the trade of essential oils and medicinal plants were the countries such as China, India, Indonesia, Brazil, Republic of Korea, USA, Chile, Egypt, Argentina, Greece, Poland, Hungary, Zaire, former Czechoslovakia and Albania, while the main markets were the countries such as European Commission, USA, Japan, Germany, France, Italy, Malaysia, Spain and UK.

The Indian Council of Forestry Research and Education (ICFRE, 2000) points out that the export scenario in India, as the trade statistics of NWFPs show, is highlighted by an increase in the foreign exchange earnings and the rate of extraction during the four decades from 1960–61 to 1990–91. The National Forestry Research Plan document of ICFRE mentions that during this period of four decades, the share of export earnings from NWFPs has been ranging between 56.5 per cent and 75 per cent of the total exports of forest produce (including both the timber and NWFPs). The trend of growth is shown more apparently by Shiva (1998) who gives evidence that the percentage share of NWFP export has ranged from 5 per cent to 15 per cent of the total foreign exchange earning that has grown from the level of ₹95 crore in 1960–61 to about ₹4,200 crore in 1990–91. The quantity (tonnage) and the value of NWFP groups exported during 1991–92 and 1996–97 also show that there is a growth of 157 per cent in the value of NWFPs exported from the country within a period of five years, while the growth in the quantity exported during the same period is 68 per cent only (Shiva, 1998).

At the international level, it is stated that 25 per cent of prescription drugs are based on plant-derived chemicals (Menon, 1996), and if we narrow down to the medicinal plant species, about 35,000 plant species are being used around the world for medicinal purposes, many of which are subjected to uncontrollable local and external trade (Lewington, 1993). WHO has listed over 21,000 plants reported to be of medicinal use throughout the world (Maudgal and Kakkar, 1996). As per another estimate, over 25 per cent of all prescription drugs in Organisation for Economic Co-operation and Development (OECD) countries and up to 60 per cent of those in Eastern Europe consist of modified higher plant products (Sharma, 1996; *The Lancet*, 1994). However, the pharmaceutical companies use not only plant extracts but also raw plant material in most of their products. There are two sources of supply of these herbals: from the wilderness, and from cultivation. It is estimated that more than 75 per cent of the world's requirement of herbal components is met from the wild (Ramakrishnappa, undated). The demand for the herbal plants is increasing in both the developing and the developed countries. However, while the demand might be increasing in the developed countries on account of the fact that the derivatives of these plants are non-narcotic, the increase in its demand in developing countries is because of the continued dependence of the indigenous people for their primary health care. As per an estimate, over 7,000–7,500 plant species found in different ecosystems are used for medicine by the indigenous people in India (AICEP, 1994; Chander, 1996). It is believed that in specialized areas such as respiratory, gastro-intestinal tract and liver disorders, skin diseases, orthopaedics, ophthalmology and mental health, the Indian traditional medical systems can make original contributions to the world of medicine (Chander, 1996). In India, the herbal industry caters to the domestic as well as international needs and has an annual turnover of about $300 million (Shankar and Majumdar, 1997). India exported 6,929 tons of medicinal plants to Germany and 806 tons to UK in the year 1982 (Lewington, 1993).

The quantum of trade in medicinal plants shown in above paragraphs, though not recent, is at least a fragment of the volume that actually crosses national boundaries. However, India undoubtedly is among the most important resource collection centres of genetic resources as well as TK. As per an estimate (Sharma, 2000), there exist 7,843 licensed manufacturers of traditional drugs in India. The practitioners of Indian system of medicines including Ayurveda, Siddha and Unani never tried for documentation or classification of the country's diversified resource of plants and animals. Similarly the issue of conservation of biodiversity also remains a matter of concern. Advocating the conservation of biodiversity that is the ultimate source of these medicinal plants, Moran (2000) states that less discussed is the vitality of biodiversity to the health of 80 per cent of the world, populations that depend solely on medicinal plants for their primary health care.

Observing the role of the native societies as regards medicinal use of plants, Moran states that of the 120 active compounds isolated from higher plants and used today in Western medicine, 74 per cent have the same therapeutic use as in native societies. She also quotes the scale of trade in herbals and gives an example that in 1997, the US dietary supplement market for herbals or botanicals was nearly $4 billion, with a $5 billion 1998 projection and compounded yearly growth rate of 15–25 per cent.

In the same context, while it may not be possible to enlist all the names or varieties of medicinal plants entering the international trade and their values, an attempt can be made to outline some of the products. For example, out of an estimated 3,000 essential oils that form an important category of the MAPs, approximately 300 are of commercial importance. The majority of these essential oils are obtained from agricultural plants, but as per UN Conference on Trade and Development (UNCTAD) assessment, approximately 26 essential oils are collected in commercial quantities from wild sources (Iqbal, 1993). In addition to the MAPs, which value in international trade around $689.9 million, some of the other valuable genetic resources in international trade (Walter, 2002) are nuts ($593.1 million), ginseng roots ($389.3 million), cork and cork products ($328.8 million) and essential oils ($312.5 million).

Based on the outflow of these products collected by indigenous communities chiefly from the wild, either in the raw or in the semi-processed/processed form, the following are the three core categories of the biological resources:

1. Biological products gathered from the wild by indigenous and local people chiefly for self-consumption;
2. Biological products gathered for making artefacts to sell in the local market in order to earn livelihood; and
3. Other mostly more valuable biological products gathered by the indigenous communities to sell to the local traders or middlemen either in raw form, or in dried/semi-processed form for better earnings and least used in self-consumption.

The first group does not involve any commercialization. The second group, involving traditional art and skill, carries some monetary gain, and the earning drawn from the products may be enhanced if the market linkage is improved. The third group comprises most of the MAPs and other high-value genetic resources that are used in manufacturing industries, often utilizing the TK associated with the resource that is held by the local communities, especially the traditional healers of the indigenous communities. The products of this group derived essentially from the MAPs are basic material for the high-valued industries producing phytopharmaceuticals, cosmetics, agrochemicals, biotechnological products and so on and enter into many new nature-based research laboratories all over the developed world because of their high commercial value and prospects. This group of products is at the focal point, especially when there is general observation that the indigenous and local communities are deprived of proper share in the benefits that accrue from them once the products enter the value-added international market. The products are exploited from the territory that is their homeland and that has been the main source of earning of their livelihoods and meeting their health-care needs over centuries, yet the indigenous and local communities are not at all compensated. When institutions and corporations draw monetary gains from exploitation of these resources, making use of the TK that belongs to these communities, particularly the traditional healers, an anomaly rises abruptly in the commercialization of these products.

Another aspect of the commercial use of MAPs relates to the personal finance that the people resort to for their use. In developed countries, pharmaceuticals are largely publicly funded through reimbursement and insurance schemes, while in developing countries, typically, 50–95 per cent of drugs are paid by the patients themselves. Thus, in developing countries, prices of medicines have direct implications for access (WHO, 2002b). In the Indian context, the domestic herbal industry is quite strong, giving rise to a reasonable expectation for significant earnings, and if the aspect of TK system developed through generations of traditional healers is taken into consideration, an enormous scope in bioprospecting comes in view. The aspect of bioprospecting has been discussed at length in the following section and is consistently mentioned in other chapters too.

Traditional Knowledge, Biodiversity Research and Bioprospecting

In the context of the commercialization of the biological diversity of tropical countries and the international trade of the components of biodiversity, an important and intricately related attribute is the TK in possession of the indigenous and local communities. These communities play an important role in the business of biodiversity—the collection of the components of biodiversity from the wild, its trade at the primary level, and providing valuable clues leading to herbal medicine and drug development through the TK they are in possession of. However, it is usual today that the present-day IPR laws are ill-equipped to seek any protection of such knowledge, and unfortunately, it is rather usual that commercial use of the TK is affecting indigenous rights and the international framework is unable to protect the rights of those who are the very source of the TK about the uses of these resources.

Some of the species that this TK is related to, particularly the species of medicinal value, are under threat because the trend of growth in demand for natural products and herbal medicines in the 'North' over the years has led to significant changes in the traditional patterns of harvesting. More and more of the favoured species, because of commercial harvesting, are vanishing from their natural ecosystem (Bhojvaid, 2003; Walter, 2002). A sense of insecurity due to this precarious nature of the earth's ecology, on the one hand, and a desire to have access to, viewed by many as economic fortune from, the South's abundant genetic resources, on the other hand, have inspired an initiative for the conservation of global biodiversity. The developing world's biodiversity, instead of being treated as common property of the indigenous communities, has in recent years been recognized as the common heritage of the whole of humankind. In response, an initiation was made by United Nations Environment Programme (UNEP) in the year 1988 that culminated into adoption of the CBD in 1992. This convention, in force from December 1993 with 194 parties to the treaty today, including India, is a big leap forward in the fair and equitable sharing of benefits from the use of genetic resources.

As regards the distribution of benefits arising out of the trade in medicinal plants from developing countries, it is observed that though the international trade in medicinal plants has increased in the past few decades with more drugs developed, little benefits accrue to the source countries and the indigenous people. According to Brown (1995), no studies have attempted to place a value on the health care provided by traditional healers and traditional plant medicines in terms of the costs of their modern equivalents, and the recent attempts to value NWFPs and in particular the medicinal plants have examined only the current local market value of these products and have not attempted any in-depth evaluation of the benefits to rural communities of traditional health strategies. She contends that in some developing countries, attempts are underway to integrate Western and indigenous medicine, and this requires the scientific evaluation of TMs, larger-scale manufacture with better quality control and training in the use of herbal remedies.

This contention depicts the growing interest in research in the medicinal plants, and policy makers and stakeholders are beginning to recognize the need for an ethical policy regarding the appropriation and use of indigenous knowledge and resources. The matter of concern is not just the unfortunate vanishing of the most favoured species from their natural ecosystem due to commercial harvesting, rather it is of equal concern that the present practices are failing to motivate young people to carry the TK forward—adding to it being negligible—for the next generation. Professor Anil K. Gupta (2005) has mentioned about it, saying:

> In last eight years, after walking more than 2600 kilometres, we did not come across many young healers or herbalists. It is obvious that available incentives and the community structures have failed to motivate young people to carry the torch of knowledge forward for the next generation.

Professor Gupta concludes that the portfolio of incentives aimed at individuals or collective can indeed reverse the knowledge erosion, and the IPRs constitute an important element of this portfolio. Karki (2003) also shows concern about the rapid rate of extinction of medicinal plants species combined with the rapid loss of indigenous knowledge systems, which has 'deep potential consequences for human health'.

Many believe an IPR system could be the best way to ameliorate this situation, allowing for appropriate financial compensation for the use of indigenous ethnobotanical knowledge. A greater understanding of this knowledge is imperative if any system of rights based upon that knowledge is to compensate the indigenous populations. Such practices would surely lead to an equitable sharing of the benefits accruing from the commercial use of biodiversity, and it is evident that biodiversity is increasingly treated now as part of a broader strategy aimed at tackling poverty, and biodiversity and well-functioning ecosystems are fundamental to both poverty reduction and sustainable development (Lapham and Livermore, 2010). Therefore, going beyond conservation and sustainable development and means of equitable benefit sharing, the biodiversity conservation concerns also attract research aspects.

The aspect is well highlighted in Article 12 (Research and Training) of the CBD that mentions that taking into account the special needs of developing countries, the CPs to the CBD shall promote and encourage research that contributes to the conservation and sustainable use of biodiversity, and shall also promote and cooperate in the use of scientific advances in biological diversity research. If we put the implications of globalization of patent laws, particularly after the advent of WTO (described in detail in Chapter 4), opposite to what is underlined by the CBD, we encounter a very complex scenario. The NGOs from developing countries talk about 'biopiracy', the CBD mandates the technology transfer, and the advancement of biotechnology poses as if the benefits that could accrue to the indigenous and local communities are at stake. Alexiades and Laird (2002) put it succinctly: The result is that complex ethical questions and challenges are made more complex by rapid technological change and globalization.

Biodiversity is an all-encompassing term,[3] and therefore the subject of biodiversity research becomes extremely wide. It comprises two major aspects. One is correlated to the wider social and national interests, and involves the assessment, survey and monitoring of biodiversity of a particular region or a well-delineated ecosystem, frequently associated with the conservation aspect too. The other major aspect of biodiversity research is correlated to the commercial interests of the researchers, and involves biotechnology, bioprospecting and drug discovery, and other purely academic research works. Both of these aspects are important, and the present trend of growing demand for natural products demands that both of the aspects must lead to the conservation of biodiversity and an equitable sharing of benefits. It is aptly iterated that although a careful balance needs to be struck between the two (the commercial interests and the wider social and national interests), important questions remain as to the

---

[3] According to the CBD, biological diversity or biodiversity means the variability among living organisms from all sources including, inter alia, terrestrial, marine and other aquatic ecosystems and the ecological complexes of which they are part; this includes diversity within species, between species and of ecosystems.

ultimate beneficiaries of these (biodiversity prospecting) agreements (Laird and Wynberg, 2002).

The requirement of striking a balance between the two may be met if the research institutions, the industries using results of such research, the funding agencies and other off-site stakeholders follow the ethics prevalent in social arena (general principles of equity and public moral with respect to the resource provider communities) and adhere to the general research guidelines (well-negotiated and equitable partnerships, transparency in sharing biodiversity information, transparency in sharing of research results). Better awareness about legal ramifications may be of greater help in striking the balance on the part of professionally accountable research institutions and researchers. Biodiversity funding also needs to be driven accordingly, focusing increasingly on social and economic objectives of equitable sharing of the benefits of biodiversity research.

It is also observed that the last decade of twentieth century has witnessed noteworthy developments in biotechnology that have opened a gamut of issues related to access to the genetic resources of the tropics and sharing of benefits. The institution of the WTO and the commencement of some of its most talked about agreements have accentuated the debate over the commercial value of biodiversity at both the national and international level. The scope of future drug discovery enhances the value of these products and the TK associated with them. The term used for this phenomenon is biodiversity prospecting or biological prospecting, in short called as 'bioprospecting'.

The Chennai (India) based M.S. Swaminathan Research Foundation defines bioprospecting as follows[4]:

> Bioprospecting is basically the search for commercially valuable biochemical and genetic resources in plants, animals and microorganisms. These resources may be used in food production, pest control, and the development of new drug and for other related biotechnological applications.

A publication of the WHO (2001a) gives a more comprehensive definition of bioprospecting:

> Bioprospecting can be defined as the systematic search for and development of new sources of chemical compounds, genes, micro-organisms,

---

[4] *Source*: www.mssrf.org/bt/bt-bioprospecting.html (Accessed in April 2014).

macro-organisms, and other valuable products from nature. It entails the search for economically valuable genetic and biochemical resources from nature. So, in brief, bioprospecting means looking for ways to commercialize biodiversity. Lately, exploration and research on indigenous knowledge related to the utilization and management of biological resources has also been included into the concept of bioprospecting. Thus, bioprospecting touches upon the conservation and sustainable use of biological resources and the rights of local and indigenous communities.

If we look at the definition of bioprospecting in the context of benefit sharing, we find that though the WIPO (2014f) defines bioprospecting in most simple words: 'The development of new therapeutics from products of nature'; this definition is followed by the concept of sharing of benefits with the indigenous communities. The WIPO report prepared by Ryan Abbott (WIPO, 2014f) mentions that bioprospecting can be beneficial to indigenous communities and developing countries if national governments and local communities may be able to receive a portion of revenue from the sale of new medicines developed from traditional resources, and this revenue can support the conservation and sustainable use of biological resources. Incidentally, both WIPO and WHO explicate that bioprospecting can be advantageous as it can generate income for developing countries, and at the same time, it can provide incentives for the conservation of biological resources and biodiversity. Surely 'bioprospecting' is an activity that may involve agreements between governments, pharmaceutical or agrochemical companies and other institutions (Hall, 2003). It includes the collection, taxonomic identification, natural history, location, field data and processing to laboratory activities (chemistry and/or biotechnology), selection of samples for screening and the final results (Mora, 1996).

The bioprospecting activity also relates to biotechnology, which is defined under Article 2 of the CBD as:

> Biotechnology means any technological application that uses biological systems, living organisms, or derivatives thereof, to make or modify products or processes for specific use.

The field of biotechnology has been witnessing pronounced activities during the last 25 years. According to Ganguli (1998), the rate of innovations in this area has been higher and their conversion to useful technologies has been rapid, making the field fiercely competitive, and protecting these innovations through patents has been a significant activity in most innovation-led companies. Referring to the growing international awareness of the link between the development of biotechnology and the preservation of genetic resources, McManis (2004) states that during the last decade of the twentieth century, two parallel rounds of international negotiations yielded two multilateral agreements—the Agreement on TRIPS and the CBD—setting the stage for further international negotiations to hammer out a more comprehensive global bargain. However, it is agreeable that no benefits would flow due to the wealth of the biodiversity of the tropical countries unless sincere efforts are undertaken for the conservation of the wealth of biodiversity.

Conservation of Biodiversity and Sustainable Use of Its Components

The need for an appropriate distribution of benefits and the global bargain raises a potential conflict between the innovators and the users of the knowledge associated with the genetic resources that provide the lead to the science of biotechnology. Quite a few of such conflicts have frequently been read during the last two decades, only to prove that it is really hard to protect the traditional indigenous knowledge under the existing IPR regime. One controversial instance of this is the case of the neem (*Azadirachta indica*) tree. Two US companies got patents for derivatives of the active principle, and the rights of the indigenous people were sidelined and they were never compensated for their TK. There are many more examples of patenting based on the TK derived from India. Rao (2001) gives the examples of biochemicals derived from *karela* (*Momordica charantia* or bitter gourd), jamun (*Syzygium cumini*) and brinjal (*Solanum melongena*) patented by a US firm for their diabetic properties, despite their uses being mentioned in several ancient Indian texts. These episodes have persistently been pursued and have been hot topics for the last several years while India and many other tropical countries have been trying to find ways as to how the new IPR regime could be made conducive to the conservation of genetic resources in the developing countries. These inconsistencies with regard to the commercial use of biodiversity and the use of TK associated with biodiversity have been studied and explained in more details in Chapter 3.

It is established that the indigenous and local communities markedly have socio-economic and cultural dependence on NWFPs (inclusive of MAPs) as well as the prerequisite of sharing of benefits accruing from these resources. This dependence, and also an equitable commercialization of these resources, necessitates their conservation and sustainable utilization. For this objective of sustainability, chiefly among the developing countries, one essential step intended could be through the provision of economic incentives. The issue of economic incentives opens the whole range of market economy, and this aspect is vital in the present context, as referring to the objectives of the CBD (conservation of biodiversity, sustainable use of its components and the fair and equitable sharing of the benefits arising out of the utilization of genetic resources), Lesser (1998) suggests that the treatment of genetic resources as the subset of genetic materials with actual or potential market value is particularly relevant, as emphasis is on the use of biodiversity, and, where needed, the creation of markets, to achieve relevant CBD objectives.

The issue very logically links the sustainability of biodiversity with its commercial use as well as domestic consumption. Referring to the limiting factors of the scope of conservation of NWFP species—genetic resources were broadly referred to as the NWFP species about 20 years ago—vis-à-vis the increase in human population in the developing countries, Myers (1995) observes that the conservation issue is closely related to the population growth, although this does not necessarily imply that more population must mean fewer species habitats. Many other factors such as inefficient agriculture, poor land use, inadequate technology, poverty and deficient trade and investment policy strategies make the picture far more complex than a simple population–biodiversity equation. As is observed by the Planning Commission of the GOI (2000), because of the pressure of population, the cultivation of food grains and commercial crops has progressively practically eliminated locally growing medicinal plants, and because of the vast changes in the social system, the family traditions have also become weak. Obviously, the factors influencing conservation of biodiversity are closely linked and yet beyond the factor of population dynamics.

The aspect of conservation of biodiversity is vital in the arena of its commercial uses. Often referred to as 'conservation biology' in educational curricula, it finds place in Article 1 (Objectives) of the CBD, more emphasized afterwards in Articles 5–13 and Article 18, pronouncing for mutual cooperation, developing national strategies, plans or programmes, identifying and monitoring components of biodiversity important for its conservation, in situ and ex situ conservation measures, education, training and developing awareness, and for international technical and scientific cooperation. Following these tenets, national biodiversity strategies and action plans have been adopted by several countries. Meanwhile the global market of the products of biodiversity, because of the enhancement in the benefits accruing from them, is apparently causing a threat to biodiversity, and if not controlled, it may lead to the extinction of endangered species and the destruction of natural habitats and resources (WHO, 2003).

It is observed that the worldwide loss of biological diversity largely continues in response to widespread habitat destruction, over-harvesting, pollution and the accidental or otherwise inappropriate introduction of foreign plants and animals (USDA, undated). Fire also is an important focus of management attention due to its effects on biodiversity (Driscoll et al., 2010). As a result, much attention is drawn to the aspect of the conservation of biological diversity today, calling for scrutiny of the efforts being made and the results of these efforts.

While considering the aspects of conservation of biological diversity, let us keep in view that biodiversity is usually treated at three levels: genetic diversity within a species, species diversity and ecosystem diversity (Kaul, 2005). At all these three levels, local communities play a key role in both conservation and selection. From the point of view of conservation of species, the assessment of status with respect to the potential or probability of adverse event (risk) of extinctions is frequently proposed as rare, endangered, vulnerable and threatened (Sharma and Negi, 2005). The International Union for Conservation of Nature (IUCN), the world's main authority on the conservation status of species, lists these categories as: extinct, extinct in the wild, critically endangered, endangered, vulnerable, near threatened, least concern, data deficient and not evaluated.[5] The IUCN Red List aims

---

[5] *Source*: http://jr.iucnredlist.org/documents/redlist_cats_crit_en.pdf (Accessed in May 2015).

to provide information and analyses on the status, trends and threats to species in order to inform and catalyze action for biodiversity conservation. However, as reported in the National Biodiversity Action Plan of India (GOI, 2008), the baseline data on species and genetic diversity and their macro- and micro-habitats are inadequate, and are not yet integrated into a national database.

Obviously, countries need to promote inter-sectoral consultations and partnerships in strengthening biodiversity conservation activities, leading to establishment of nationwide information system with uniform format for collection, retrieval and dissemination of data. A suitable conservation strategy is difficult option in view of socio-ecological factors that may constrain the choice of available strategies (Kumaraswamy and Kunte, 2013). For example, certain conservation programmes that address food security needs will often invest in different places and ways than those focused on saving species from extinction (Kaimowitz and Sheil, 2010).

One of the major causes for the loss of biodiversity is forest fragmentation leading to isolated forest patches created due to changes in local land use, which further leads to the extinction of local forest species. Fragmented forests will contain fewer number of the original forest species compared to what the continuous forests would contain, and efficient biodiversity conservation programmes would require an ability to predict the scale of losses of biodiversity (Zuidema et al., 2005). The limited financial resources available for conservation programmes, especially in developing tropical countries, need to be used efficiently, and although socio-economic constraints might limit its success, the establishment of extensive protected forest areas is suggested in many conservation plans (Zuidema et al., 2005).

Narrating about two competing paradigms for forests—a paradigm of segregation between production and protection functions and another of integration, a traditional approach where near-natural forests are managed carefully to yield timber and other products and simultaneously to maximize biodiversity and other environmental values—Sayer et al. (2005) suggest a third paradigm of optimizing outcomes through landscape mosaics. Under this paradigm, some areas of special significance for biological diversity must be given total protection as inviolate nature reserves. Additionally, we will need extensive semi-natural forest to provide the matrix between protected areas, to sequester carbon and regulate hydrological functions, along with a process to ensure that the different elements of the mosaic are complementary, for example, habit corridors that are continuous between protected areas, and must actually provide free species movements. In addition, the mosaic should include watershed protection forests (parallel to contours) and intensively managed plantations so as to minimize opportunity costs for agriculture.

Apart from the concept of establishment of extensive protected areas suggested for conservation of biological diversity, and quite different from the concept of landscape mosaics, India's Biological Diversity Act, under Section 37, has a new concept of establishment of Biodiversity Heritage Sites (BHS). These are the well-defined areas of biodiversity importance established by the Government, aiming to stem the rapid loss of biodiversity. A few significant features of BHS concept (NBA, undated) are:

- The BHS should be unique, ecologically fragile ecosystems. The sites that are like existing sacred groves in general can be declared and notified as BHS. The sites that are not covered under protected area (national park and sanctuary) network may be considered.
- In consultation with the local bodies, the Government (the state government or GOI) may notify the areas of biodiversity importance as BHS. The management of the BHS shall be the responsibility of the local Biodiversity Management Committee (BMC) or the BHS Management Committee (in case of the BHS overlapping on more than one BMC).
- A management plan for each BHS shall be prepared with a map of the notified administrative boundaries, the status of ownership, major components of biodiversity, past and current land use pattern, management prescriptions and projection of the expected outcomes.
- Any programme or activity implemented by the government or any other agency that is likely to have adverse impact on the BHS may be avoided.
- The creation of BHS is not to put restrictions on the prevailing practices of and usages by the local communities, except those voluntarily decided by them. The purpose is to enhance the quality of life of the local communities through conservation measures.

The Biological Diversity Act, under Section 37, also mentions that the state government, in consultation with the GOI, may frame rules for the management and conservation of BHS, and the state government shall also frame schemes for compensating the section of people economically affected by such notification.

In the context of sustainability, Sheldon and Balick (1995) state that since most industries have more experience with marketing than with determining levels of sustainability, levels of sustainability are often developed more according to levels of demand than to actual population dynamics of a natural supply, and it is probable that many more so-called 'green' products are sold than could potentially be 'sustainably' harvested. Swanson (1995) observes that the solution to the global biodiversity problem requires the creation of some mechanism for appropriating the values of 'evolution-supplied information'. He concludes that the conservation of diversity and diversity's values will require the creation of a diverse set of international institutions. Referring to the two major threats to medicinal plants, Brown (1995) states that the first is the loss of habitat (through land use conversion, agricultural expansion, etc.) and second is the overexploitation of known species as a result of increased demand, and related to these two is the associated loss of indigenous knowledge and expertise. Highlighting the conflicts related to the conservation of medicinal plants from the perspective of the developing world, Khalil (1995) opines that the ownership of genetic resources and, indeed, of medicinal plants is likely to precipitate major tensions between biotechnology firms and the developing countries, particularly where such resources have been directly conserved and utilized by specific indigenous communities.

Alluding to a more practical approach, Sarin (2003b) states that agro-technologies for cultivation of a number of plants have been developed but a majority of these have remained unutilized. He observes that this calls for development of appropriate technologies that should be in line with the agricultural practices used by the farmer and reduction in the cost of cultivated produce, since the cultivation of medicinal plants failed largely on account of absence of market demand while the drug industry continued to get almost 90 per cent of its supplies from the collections made from the wild. Kuipers (1997) reiterates that the problem lies in the fact that there is little or no legislation restricting the use of wild-harvested materials in finished products, or for assuring the sustainable utilization of medicinal plants. Elaborating on this issue, Walter (2002) states that the international trade in NWFPs leads to high market value compared to local or national markets, and this in turn may lead to the overexploitation (and ultimate extinction) of certain species. In addition to this, several other factors such as population growth, growth in the demand for nature-based products (cosmetics, pharmaceuticals and other industrial products) and so on trigger overexploitation. The high market value often exhorts the indigenous and local people to gather and sell in the market for cash earning the produce that they would otherwise have retained for self-consumption. It is often argued that the concept of economic incentive can take care of such anomalies if provided through a well-laid-out programme.

This further leads to the basics that the resource managers need to focus on. Evidently, the issue of the sustainability of biodiversity is very closely linked to its sustainable commercial use. Proper harvesting techniques, better resource uses and augmentation with prioritizing the species and cultivation are essential along with an equitable sharing of the benefits accruing from the resource. The efforts for conservation will be useful if accompanied with economic benefits, and sustainability may be possible through cultivation and scientific management and the use of appropriate technology for nursery and plantation, which is available yet used inadequately. Sarin (2003a) validates that large-scale cultivation of medicinal plants, both indigenous and exotics, will be a desirable solution for ensuring unrestricted supply of the raw materials in required quantities.

The Task Force formed by the GOI on conservation and sustainable use of medicinal plants identified 25 species that were in great demand in the herbal medicine (GOI, 2000), and recommended for improving awareness and availability of these species as well as their cultivation through involvement of NGOs. These species were *Aconitum heterophyllum* (Indian Atees), *Aegle marmelos* (Bael), *Andrographis paniculata* (Kalmegh), *Asparagus racemosus* (Shatavari), *Bacopa monnieri* (Brahmi), *Berberis aristata* (Indian Berberry), *Cassia angustifolia* (Senna), *Chlorophytum arundinaceum* or *A. borivilianum* (Safed Musli), *Commiphora wightii* (Guggal), *Convolvulus pluricaulis* (Shankhpushpi), *Embelia ribes* (Baiberang or Vai Vidang), *Emblica officinalis* (Amla), *Garcinia indica* (Kokum), *Glycyrrhiza glabra* (Liquorice or Mulethi), *Gymnema sylvestre* (Gudmar or Madhunashini),

Nardostachys jatamansi (Jatamansi), *Picrorhiza kurroa* (Kutki), *Piper longum* (Long pepper or Pippali), *Plantago ovata* (Isabgol), *Santalum album* (Chandan), *Saraca asoca* (Ashoka), *Saussurea costus* or *S. lappa* (Kuth), *Swertia chirata* (Chirata), *Tinospora cordifolia* (Giloy) and *Withania somnifera* (Ashwagandha).

The Task Force also recommended that in order to fully convert the potential of medicinal plants into economic wealth, a very active R&D programme is essential, which should focus on the following five points:

1. Evolving and optimizing the most appropriate technologies for conservation, especially for endangered or endemic species and molecular methods for characterization;
2. Detailed studies on life cycle and breeding behaviour, taxonomy, seed biology;
3. Population and habitat viability studies;
4. Optimizing appropriate methods for post-harvest handling, processing and storage; and
5. Investigation on quality control standardization and shelf life of raw materials and finished products.

The approaches stated above need careful investigation as to what type of scheme can ensure the most equitable distribution of economic incentives. In fact, an approach to encourage the conservation of biodiversity through the provision of economic incentives is now part of many accords related to the transaction of genetic resources and associated TK. The exemplary collaboration between the INBio of Costa Rica and Merck & Co. of the USA, a case of advance payment of the cost of genetic resource, has plainly shown this approach, though the people were not a part of the accord at all (Dutfield, 1997). Indigenous communities in the tropics have continued access to their local natural resources as a right. No doubt that the state is the ultimate owner of all resources and the ultimate right has to apply to the state, yet the indigenous communities might be enabled and empowered so that they have full usufructs over the biodiversity that is situated in their geographical proximity. Therefore, as a natural corollary, the indigenous people's right to self-determination has got to be recognized before they are prepared to enter into negotiations over access to their resources. The CBD has a provision for equitable sharing of results of R&D and the benefits arising from the commercial utilization of genetic resources. There is a need to actualize this provision, and this calls for a wholesome study of the existing legal mechanism in order to safeguard the rights of the indigenous people.

The advent of biotechnology and the global uniformity of intellectual property (IP) standards are issues that also need to be pondered as these are likely to impinge on sustainability issues in general. It may be noted that despite its potential gain and despite several contractual programmes taken up the world over, few indigenous communities have benefited from bioprospecting. The argument that access to tropical genetic resources, if denied, would limit innovation, and at the same time unrestricted access would trigger their destruction and extinction, is apparently valid. What is required is to look for a mechanism for regulating the access so that the ambitions of the developed and also the developing nations can be achieved simultaneously.

It is possible to meet this requirement after an analytical review of the present situation of commercial use of biodiversity at local, national and international level, with the aim to bring out the bottlenecks, including those with respect to equitable sharing of the benefits. The complete range of the commercial use and sustainability of biodiversity includes the happenings of the last decade in the form of international agreements and conventions such as the Agreement on TRIPS and the CBD with special reference to the commercialization and equitable benefit sharing in the tropical regions of the world. The deliberations held at national and international platforms during the recent past have identified some important aspects such as exploitation of genetic resources of developing countries by developed countries, safeguarding the TK of indigenous people and the rights of stakeholders and have warranted a careful investigation of the related subject.

Over the past two decades, a set of other national and international institutional arrangements have come up to circumvent the bottlenecks discussed above. A lot of new information has been generated on the potential of commercial use of biodiversity, its socio-economic importance and benefit-sharing mechanism. Quite a few institutions (governmental, non-governmental as well as cooperative) and the private sector have become involved with the trade and manufacture related to genetic resources, and the developments ought to be persistently analysed in order to derive a model mechanism of more equitable sharing of benefits. The juridical developments such as India's Biological Diversity Act of 2002, the Rules made there under, the framework of ABS laws being formulated in many countries and the changes in patent laws the world over after the advent of the WTO also need to be evaluated as to how far these instruments are able to ameliorate the situation.

To sum up, it may be restated that the biological resources from the natural forests in developing countries have provided people with a rich supply of food and other basic needs of life since ages. The components of these resources are gathered for consumption and also to sell in the local weekly market. On the assessment of the status of biodiversity attempted in this chapter, it is recognized that the indigenous populations geographically overlap with the forest areas in a country, and sadly enough, the populations in and around the forests are the laggards in the socio-economic development process, living in primitive conditions. It is also observed that a wide range of the products of biodiversity leads to their use in a wide extent of markets at the national and international levels, though the data related to this trade are too scanty. The available data are clearly indicative of the general flow of these products from the biodiversity-rich and technology-poor 'South' to the biodiversity-poor and technology-rich 'North', and a well-demarcated North–South divide is seen. The aspect of bioprospecting also is creating a challenging situation as it is realized that little benefits accrue to the source countries and the indigenous people. The issue of the sustainability of biodiversity is very closely linked to its sustainable commercial use. Simultaneously, the growing demand for nature-based products in the international market and the attention of transnational corporations towards indigenous genetic resources of the developing world also need to be deliberated upon. A proper evaluation of the approach adopted by the tropical countries at the policy level is required, and suitable mechanism is necessary to earn appropriate returns from the wealth of the biodiversity and to make sure equitable sharing of the benefits among the various on-site and off-site stakeholders.

3

Traditional Knowledge in the New IPR Regime

The most contentious issue in the context of access to genetic resources and the sharing of benefits accruing from them has been the one related to the TK associated with these resources that has been held by the indigenous and local communities. An acceptable solution for the protection and promotion of TK has yet to emerge despite the endless debates during the last 20 years and the unconstrained efforts made for protecting the TK under the new IPR regime after the advent of the WTO. A large number of bodies including the WIPO, the SCBD, the UNCTAD and even the WTO are engaged in the discussions on the protection aspect of TK. The issue is the subject matter of enormous campaigns run by numerous NGOs the world over and covers incessant debates, workshops and other kinds of activism. Side by side, during the period of the last 10–15 years, many tropical countries either have already formulated or are in the process of formulating laws for regulating access to the genetic resources and the sharing of benefits accruing from the commercial use of these resources and the TK associated with them.

Before reviewing the state of affairs with regard to this subject area, it would be pertinent to look into the essential elements that would proffer a clear-cut perception about the TK and the different facets of the viewpoint as to why it is an essentiality to share the benefits arising out the use of TK. Johnson (1992) defines the Traditional Environmental Knowledge as:

A body of knowledge built by a group of people through generations of living in close contact with nature. It includes a system of classification, a set of empirical observations about the local environment, and a system of self-management that governs resource use.

This definition defines the TK excellently in the background of the commercial use of genetic resources, as it includes all the aspects that form a prominent base with respect to the protection of TK in the realm of the new IPR regime and the need for an appropriate sharing of benefits drawn out of the use of the body of TK and the genetic resources that are part of the habitat of the indigenous and local communities. More explicitly, the TK is understood to be a corpus of empirical knowledge built up through generations that governs resource use, is recorded and transmitted through oral traditions, learned through observation and hands-on experience, and is checked, validated and revised daily and seasonally through the annual cycle of activities (Johnson, 1992). Dr Vandana Shiva (2001), who is known for her movement for the protection of the diversity and integrity of living resources, especially native seed, states that the indigenous (traditional) knowledge is centred on co-creation by the nature and the people. An appropriate definition of TK propounded by the Sector Paper on Traditional Knowledge of Namibia National Biodiversity Programme (NNBP, 1999), in terms of biodiversity, in agreement to the above definition and consistent with Article 8(j) (*Equitable sharing of the benefits of traditional knowledge*) of the CBD, is as follows:

The knowledge, innovations and practices of indigenous and local communities embodying traditional lifestyles relevant for the conservation and sustainable use of biological diversity.

The TK is thus articulated as a means that embodies traditional lifestyles of the local communities evolving indigenously over time, and it is stated that the body of TK is inherently complex, varied and dynamic in its shape and substance (Balick, 2003; Gupta, 2002b; Kiene, 2011; WIPO, 2010). However, it is noteworthy that the CBD (Article 2: Use of Terms) does not incorporate any definition of TK, though other important terms related to the commercial use of biodiversity get mentioned in it. The term of TK, although not literally, is included in its true sense in Article 8(j) (*Equitable sharing of the benefits of traditional knowledge*) of the CBD. The TK is not defined in India's Biological Diversity Act too, though the Act alludes to the TK in the definition of 'benefit claimers'.

Despite these nuances, the significance of TK is not lessened at all, and this term is continually made a note of and is the central idea in all the discussions about the commercial use of biodiversity. Even so, the debate on the adequacy and ethics of IP protection, particularly the absence of consensus on whether and how to extend IP protection to TK, has so far shown that the issue of IP protection of TK is complex and controversial. The premise revolves around the policy related to the modalities of such protection, its implication for the indigenous people and the ownership and enforcement of the rights. The deliberations in the following paragraphs encompass most of these facets related to the protection of TK.

Vis-à-vis the aspect of its protection in the present-day IPR regime, TK held by the indigenous and local communities consists of the collective innovation that is altogether different from any individual innovation that is localized in time and space. Dr Vandana Shiva (2001) notes that this collective innovation is modified and enhanced as it is used over time and passed on from generation to generation, and in some examples (e.g., in the case of seeds and in the case of major knowledge traditions such as Ayurvedic and Chinese medicine), it is no longer local, and in some cases, it even crosses national boundaries. In order to enhance the scope further for better understanding, it would be apt to refer to Charles R. McManis (2004), Professor of Law at the Washington University, St Louis, USA, who says that some collective knowledge of a community corresponds more closely to what in Western cultural and legal terms might be called corporate, proprietary or closely held know-how, and would thus be capable of either being protected as proprietary know-how or shared with the rest of humanity, depending on the consensus of the collectivity that possesses it. However, the reality is more complex. The research scholars in several universities who visit the field to study the trees and plant products and interview the members of indigenous and local communities about the uses of the plant resources come back and write their theses and dissertations describing a wealth of ethnobotanical information. Now if this information is screened, analysed, collated and made use of, covering up the whole into a single pocket or single body of knowledge after an intricate process of research that is not confined to an individual researcher and after years of innovation, a patent is granted in some corner of the world far away from the indigenous and local community that since ages has been holding the TK associated with the research area under reference, then the question is: How to formulate, in such a scenario, the framework to protect that TK and share the benefits that would flow in?

Another depressing account is that the knowledge so generated is bearing the risk of disappearance in future. Singh et al. (2012) observe in a study conducted in Nepal that indigenous knowledge on usages of medicinal plants is transmitted without any systematic process, and younger generations of the tribes are not interested in traditional healing system because it has very little scope for money, so they engage themselves in other occupations. Murad et al. (2013) also observe, in a study in Pakistan, that ethnomedicinal knowledge is becoming restricted only to the elders, traditional practitioners and local herb sellers, and advancement in science and technology is changing the social values, therefore younger generation is transforming at a much faster rate into the new tradition, and as a result, medicinal plants knowledge is going to be obsolete. It is also true that the indigenous and local communities, who are not only unacknowledged for the information they have shared about the various uses of biodiversity and the TK associated with these resources, are helpless if there is any claim over the benefits that are drawn when their TK is made use of commercially, merely on the ground that the information is now in the public domain.

The reason is that many of the research findings get published in scientific journals and books. It is common knowledge—the vast number of books and papers published around the world on the medicinal uses of plants illustrates how simple it is to copy medicinal TK. Quite obviously, the relevance of TK as a useful source of information for researchers in the pharmaceutical field who seek to identify new chemical and biological elements, as well as new approaches to disease treatments, is generally undisputed (Carvalho, 2003). It is also observed that discovery of major part of the plant-derived drugs resulted from the active substances isolated from plants that were used originally in TM, and that the scientists did not hesitate in drawing upon the useful, valid and abstractable local knowledge when it was appropriate (Farnsworth et al., 1985; Gupta, 2001). It is also recognized that TK has been central to the modern drug development as the hit-rate has been higher in the case of TK-based research as compared to that in the random search methods in pharmaceutical industries, and even the costs reduced considerably (Gupta, 1999a).

Fortunately, the significance of the TK is being well accepted—it is assumed to be essential to the food security and health of millions of people in the developing world (CIPR, 2002), and a fact sheet of WHO (2002a) rightly records that in many countries, TMs provide the only affordable treatment available to poor people, and in the developing countries, up to 80 per cent of the population depend on TMs to help meet their primary health-care needs (Farnsworth, 1988). The TK associated with the biological resources has been playing foremost part in the lifestyle of the populations of developing countries who chiefly depended on these species not only for their primary health care but also for their nutritional needs and income generation.

Another noteworthy aspect about TK is that it is being created every day, evolving as a response of local communities to the challenges posed by their social environment (WIPO, 2003b), and accumulated over millennia, it is as rich and diverse as tropical forests' biological resources, and as threatened (Moran, 2000). Interestingly, as Watson and Gamage (1998) also assert, that there is considerable demand for TK, created by the international market, related to conservation and sustainable use of medicinal plants, including ethnobotanical and ethnopharmacological information, as compared to TK in other fields. This is the reason why the argument is gaining ground for appropriate measures to be provided for the protection of TK. Multitudinous options and alternatives are suggested by different authors for the purpose, a few of which are examined in the following section.

Quest for the Mode of Protection of Traditional Knowledge

There is a strongly advocated and quite broadly accepted suggestion that the present IPR regime has to be reformed in order to protect the TK. Utkarsh Ghate, who runs an NGO for rural environmental development in Chhattisgarh, states (Ghate et al., 1999) that the purpose of these suggestions is to ensure that folk knowledge and resources too have a share in the commercial benefits, since traditionally the IPR legislation is only meant to protect the ability of an entrepreneur to monopolize the market and corner the benefits. Ghate et al. (1999) further state that the IPR laws cannot on their own provide for sharing of the commercial benefits with the public domain foundation of knowledge and resources underlying the innovation—that function, in India, is being entrusted to the NBA, which is being constituted in India under the Biological Diversity Act. This Act of 2002, which came into force in February 2003, has been appraised in Chapter 5.

As regards the instruments that could be employed for protection of TK, Professor Anil K. Gupta (2001) rightly states that we should not assume that every bit of knowledge, innovation and practice can be protected by the same instrument. The most commonly used legal instrument to protect the right to benefit financially from scientific innovations is the patent law. Nevertheless, as practised in most countries, it is an inadequate tool to provide for sharing of the benefits from bioprospecting as the patent law is generally unable to recognize stewardship of biodiversity or maintenance of TK of the uses of biodiversity. In recent years, the contractual agreements among bioprospecting partnerships have been more widely considered as a better means of securing benefits for the resource provider country.

However, it would be evident from the review of the case studies of the bioprospecting partnership programmes recounted in Chapter 6 that these programmes have to be very well funded if the patent mode is the option for protection of TK. Patent itself has a limitation with respect to the protection of IPRs, as the process of obtaining and maintaining patents is a costly affair. The actual cost includes the fee for application, renewal charges, defending against infringements and, above all, the multiplicity of requirement of protection that may include a number of patents, trademarks or copyright for a single new product. Ganguli (1998) states that to retain competitiveness, one will have to conceive of a number of forms of the same or structurally similar molecules that have the potential of acting as a drug, and protect them through a judicious set of broad claims in a number of patents. Now if there is an expectation that a self-help group or an indigenous community might opt for the patent mode to protect the

Box 3.1:
The Biodiversity Laws of Peru

Peru has since enacted its major Biodiversity Law in 2004, known as 'the Act on the Protection of Access to Peruvian Biological Diversity and the Collective Knowledge of Indigenous Peoples' (Law No. 28216), though few more laws and regulations coexist in Peru, governing the biodiversity, access to the genetic resources, native breeds and the collective knowledge of indigenous people. These are: (1) Protection Regime for the Collective Knowledge of Indigenous Peoples derived from Biological Resources, 2002 (Law No. 27811); (2) Law declaring Crops, Native Breeds and Usufruct Wildlife Species part of the Nation's Natural Heritage, 2005 (Law No. 28477); (3) Resolution approving the Complementary Provisions to Decision 486 of the Andean Community Commission establishing the Common Regime on Industrial Property, 2008 (Legislative Decree No. 1075); and (4) Regulation on Access to Genetic Resources, 2009 (Ministerial Resolution No. 087-2008-MINAM).

Source: www.wipo.int/wipolex/en/results.jsp (Accessed in November 2014).

TK that it holds, even with usufructs over the resource that may be endemic to its locality, it would be far from reality and that is the reason the tropical countries are still struggling to find a way out.

Certainly to focus all efforts in protecting the TK through the prevalent IPR mode will imply very high transaction costs, making the effort too expensive and even ineffective. Suggesting to have 'good corporate practices' or institutional codes of conduct in this connection, Ruiz (2002) cites the example of a draft regulation in Peru, which incorporates this approach and seeks to create an incentive so that a potential user of TK in the public domain at least considers the possibility of negotiating with communities over this knowledge (see Box 3.1). Ruiz considers that TK that is in the public domain—especially when the knowledge has surpassed the physical and geographical boundaries of communities—could be hard, but not impossible to protect in positive terms.

Apparently, despite a great deal of optimism shown by authors like Ruiz, ensuring the protection to TK is not a cakewalk. McManis (2004) observes that proponents of strong IPRs all too often tend to dismiss efforts to create stronger legal protection for the TK of indigenous communities as 'the laughable proposals of confused romantics', in as much as TK is essentially static and non-new, thus having no legitimate claim to IP protection and being of marginal utility at best in the innovative process. It may be stated in this context that this is all a matter of perception, as the very nature of TK is dynamic as elucidated in the above section. Even many tangential issues such as the hitch of collective ownership and enforcement of rights also need to be resolved before a regime for the protection of TK is developed (Correa, 2001). In this context, Mohamed Khalil (1995) states that the use of property rights as incentives to conserve medicinal plants has been recommended as an important policy option, though the approach of the West to the issue of property is too foreign to the way indigenous communities view the same. He suggests that what is needed by the world is not uniform IPR system, but a diverse one that respects the rights of traditional cultures. Singh et al. (2000) observe that considering that most centralized governments, especially in the countries of the 'South', have miserably failed to safeguard such access or even to provide viable alternatives, it is difficult to distinguish between the right to survival and the right to local resources. For sure, the policy of restricting access to the natural resources has to be rational and an arrangement needs to be made in order to devise a mechanism for meeting out the requirements of earning livelihood and fulfilling the health-care system of these people through sharing of the benefits accruing from the biological resource that is situated in their geographical proximity.

Deliberating on this issue, Patnaik (1997) asserts that while for the rich countries, IPRs are fundamental rights comparable to the right to physical property, for the poor countries, protection of IP is viewed fundamentally as an economic policy question. He affirms that the inclusion of IP consideration in the Uruguay Round negotiations of General Agreement on Tariffs and Trade (GATT) as one of the new themes was unpalatable to the developing countries, and there are some misapprehension among the business elite in the developing countries of an impending takeover by the transnational corporations. Discussing this aspect, Carvalho (2003) suggests that in order to make whatever regime effective, it is of essence to promote the participation of the holders of TK in national and international discussions. In essence, it is observed that only with a deeper understanding and greater practical experience at national or regional level would it be realistic to develop an international system for the protection of TK. Consequently, it is most desirable to protect TK as a valuable corpus of IP, and as this corpus of knowledge is found as too vulnerable to be protected with the help of the framework of the Agreement on TRIPS, new strategies have to be found for its protection.

As has been alluded to in the beginning of this chapter, a large number of bodies including WIPO and UNCTAD are conferring in this direction with the principal view of evolving mechanism for the protection of TK. Under its Plan of Action, adopted at its Bangkok session in 2000, UNCTAD was directed by member states, taking into account the objectives and provisions of the CBD and the Agreement on TRIPS, to study ways to protect TK, innovations and practices of local and indigenous communities (Carvalho, 2003). It is contested to envisage a system that might provide legal protection to TK and simultaneously contribute to the promotion and dissemination of innovations based on TK. The strategy for protection might have dimensions pertaining to the community, national, regional and international levels; the stronger the integration and coordination between these dimensions, the more likely the overall effectiveness would be (WIPO, 2003b). But still the situation is not so promising, and it will be examined in the following paragraphs how the developed world is behaving as regards the protection of TK and what are the factors that are responsible, in the context of the new IPR regime that came into being along with the WTO, for rendering the protection aspect of TK at great risk.

Examining the Agreement on TRIPS critically with a view to draw support for the protection of TK, Professor McManis (2004) argues that any TK that is or might become widely known, thus losing any claim to protection under Article 39 (*Protection of Undisclosed Information*) of the Agreement on TRIPS, may qualify as one or more of the following forms of IP for which legal protection is mandated by the Agreement on TRIPS:

1. *Expression of traditional knowledge and folklore as literary or artistic works or industrial designs*: When members of indigenous or local communities provide expressions of TK or folklore to researchers, they must be recognized as the author of that

work. TRIPS itself adopts Article 7*bis* of the Berne Convention,[1] which explicitly recognizes the concept of joint authorship.

2. *Expressions of traditional knowledge or folklore as trademarks, certification and collective marks, and/or geographical indications*: While invoking the form of protection for particular expressions of TK or folklore may seem to encourage or even require the 'commodification' of culture as a condition of qualifying for protection, this is not necessarily the case. The Agreement on TRIPS itself requires (*Article 22: Protection of Geographical Indications*) that WTO members protect any geographical indication that identifies a good as originating in a particular territory, region or locality of a member, where a given quality, reputation or other characteristic of the good is essentially attributable to its geographical origin.

3. *Traditional knowledge as can be protected under the law of patents or plant variety protection*: Article 27 *(Patentable Subject Matter)* specifies that a WTO member must provide for the protection of plant varieties either by patents or by an effective sui generis system or by any combination thereof. Both sides in the debate over TK protection tend summarily to dismiss the availability of patent protection for TK, and for remarkably similar reasons.

Eventually, despite all the factors discussed above that seek to support the protection of TK to a great length, we find that the problem largely is too complex, and there are many other limitations in the protection of TK. Let us examine this problem further.

Biopiracy and Other Limitations in the Protection of TK

The new concepts and novel ideas brought up during the recent past for understanding the intricacies involved as regards the protection of TK have been illustrated in the above section. What if the TK as such

---

[1] Berne Convention for the Protection of Literary and Artistic Works, 1886, is a WIPO-administered treaty dealing with the protection of works and the rights of their authors. It is based on three basic principles and contains a series of provisions determining the minimum protection to be granted, as well as special provisions available to developing countries that want to make use of them. *Source*: www.wipo.int/treaties/en/ip/berne (Accessed in November 2014).

does not get the protection it deserves and the indigenous and local communities are deprived of the benefits they claim they could get when the TK was used by pharmaceutical corporations for commercial gains? When these issues are questioned and are inquired deeper, what comes out is that the globalization set off with the beginning of the WTO, and its new-found IPR regime leaves no option for the developing countries except going for the reconciliation between the Agreement on TRIPS and the CBD, so that benefits could be drawn and shared with the indigenous and local communities in general and the holders of the TK in particular. The advantage shall have to be drawn from both of these instruments, and how this reconciliation between them is achieved is still unresolved, whereas there are no evidences to confirm that the global misappropriation of the TK held by the indigenous communities have been brought to a halt.

The strife arising out of the misappropriation mentioned above has been referred to frequently, and in the recent past, several cases of obtaining patents making use of the TK have been recorded, reporting the fear of 'biopiracy' as a reality. The term of biopiracy, initially used by NGOs, is defined in the American Heritage Dictionary as follows:

> The commercial development of naturally occurring biological materials, such as plant substances or genetic cell lines, by a technologically advanced country or organisation without fair compensation to the peoples or nations in whose territory the materials were originally discovered.

Webster Online Dictionary[2] defines 'biopiracy' as biological theft or illegal collection of indigenous plants by corporations who patent them for their own use. A more comprehensive remark on biopiracy is given by the Erosion, Technology and Concentration (ETC) Group[3] as shown below:

> Biopiracy refers to the appropriation of the knowledge and genetic resources of farming and indigenous communities by individuals or institutions who seek exclusive monopoly control (patents or intellectual

---

[2] *Source*: www.webster-dictionary.org
[3] *Source*: etcgroup.org (Accessed in October 2014). The ETC Group, known as Rural Advancement Foundation International (RAFI) until 2001, is an international organization dedicated to 'the conservation and sustainable advancement of cultural and ecological diversity and human rights'. The full legal name is Action Group on Erosion, Technology and Concentration.

property) over these resources and knowledge. ETC Group believes that intellectual property is predatory on the rights and knowledge of farming communities and indigenous peoples.

Arguing that the globalization of IPRs is promoting biopiracy, Dr Vandana Shiva (2000) emphasizes that there are certain distortions in the US law that facilitate the patenting process for companies. Referring to one such distortion as the interpretation of *prior art* that permits patents to be filed on discoveries made in the USA, whether or not identical ones already exist and are in use in other parts of the world, Shiva reiterates that unless this part of the US patent law—Section 102—is amended, new examples of biopiracy will continue to occur. Addressing to the issue of patents granted in the developed countries based on the TK originating in India, Rao (2001) claims that RiceTec, a US firm, was granted patent for inventing a rice line with characteristics similar to those of basmati, on the specious plea that it had developed a new strain of rice. Similarly, WR Grace & Co. was granted patent on the fungicidal properties of neem, though known in India for ages, which was challenged when Navdanya (India-based NGO) started the campaign against biopiracy with the Neem Campaign in 1994 and filed a legal opposition against the USDA and WR Grace against neem patents.

Stating the examples of biochemicals derived from *karela*, jamun and brinjal patented by a US firm for their diabetic properties, despite their uses being mentioned in several ancient Indian texts, Rao (2001) further states that the lacuna is that in much of the Third World, there is no meticulous recording of information available in the public domain. The Indian system of medicine vis-à-vis modern medicine is not strong on issues of standardization. The time seems to have come when India, and so also other tropical countries, heeded the warning signals and got its act together by systematically preparing an inventory of all such knowledge, record it and put it in the public domain. That has been the most commonplace belief, and the process has begun as regards documentation of the TK in India.

The recent examples of misappropriation of TK-based innovations mentioned above, such as neem and others, which had made news headlines during the late 1990s, are an indication of the vulnerability of TK that belongs to the developing world at the hands of the developed world. The economic loss due to this misappropriation has been assessed under a UN study (Nair, 2002b): If royalty was paid by the developed countries to the developing countries for use of their knowledge base and natural assets, the latter would have gained as much as $5 billion, double the amount claimed by the developed countries as loss due to patent piracy. Another example that shows the vulnerability of TK is that the US law, in granting patents, takes account of US oral traditions and knowledge (in terms of what is already public and not patentable), but does not recognize such TK and oral traditions abroad, but only material 'published' in scientific or other journals (Raja, 2001). The developing countries have been following the 'first to file' system for accepting the claim of patents, while the 'first to invent' system, as used in the USA, is far more favourable to the small, scattered and disadvantaged innovators (Gupta, 2001), who remain unrewarded when their knowledge is exploited commercially by third parties. This weakness of TK-holders is attributed by the WIPO (2003b) to the lack of unified voice and lack of clear national policy concerning the utilization of TK.

While the UN institutions are carrying on extensive steps, and WIPO also on its own undertaking extensive work on IP aspects pertaining to the protection of TK and ABS for associated genetic resources, the outcome is still too far. The mechanism being followed in the process draws substantively from the tenets of the CBD, while Nair (2002b) reports in this context, quoting about the consensus at a meeting organized by the GOI and UNCTAD, that the system under the CBD or the Agreement on TRIPS is not adequate or appropriate for protection of community rights on TK and indigenous systems and practices.

The reasons why the existing IPR regime is supposed to be inappropriate for the protection of TK have been identified and listed by numerous other authors. These reasons may comprehensively be listed as shown below:

1. The industrial research is viewed as an individual activity, in a fixed time frame, that comprises specific ideas and identifiable components, capable of description in a way that can be owned by an individual, and thus patented easily under the existing IPR regime. In contrast, most of the TK results from sharing new ideas, knowledge and experiences shared between people and passing through generations, spanning over an unlimited time fame, in a way that it cannot be owned by an individual or a group.

2. Industrial researchers can patent TK-based products and processes through intensive research efforts that they are capable of, making distinctive improvements in the potency and efficacy of the final products that validates non-sharing of the benefits accruing from the very knowledge that provided the lead to those products.

3. The holders of TK in general lack the sensitive commercial information required to work in an IPR regime, and are mostly unequipped to handle their knowledge base and use of the resource in a confidential manner.

4. The filing, defending and maintaining of patents is a costly affair, which is far beyond the capacity and means of the holders of TK, of course, unless they are helped through some capacity-building programme.

A study conducted by the Research and Information System for Developing Countries (RIS, 2003), however, suggests that the developed countries do not in general contest the rights of countries to protect the TK (see Box 3.2). They generally hold the view that the

Box 3.2:
Research and Information System (RIS) for Developing Countries

The RIS, a New Delhi-based autonomous think tank under the Ministry of External Affairs, GOI, specializes in policy research on international economic issues and development cooperation. RIS is envisioned as a forum for fostering effective policy dialogue and capacity building among developing countries on international economic issues. As part of this research programme on promoting South–South Cooperation, RIS has taken the major initiative to bring academics, civil society organizations and the policy makers on one platform. It also seeks to strengthen policy coherence on international economic issues. The institute has launched a number of studies as part of its research activities in the area of promoting South–South Cooperation.

Source: www.ris.org.in/about-ris/about-the-institute.html (Accessed in November 2014).

Agreement on TRIPS and the CBD are mutually supportive and object to the idea of disclosure requirement of genetic resources in the process of patent applications. Some developed countries such as the USA, European Union (EU), Japan and Norway have said that the discussion in the Council for TRIPS should wait for results of the work being done at various international forums, such as the WIPO.

Against this backdrop, now an attempt is made in the following section to look for the options that are available for the protection of TK and, most importantly, the options available for claiming the benefits that accrue to the members of the indigenous and local communities when the TK is used by the pharmaceutical, cosmetic and other industries that are in the business of natural products.

Options for Protection of TK

An effort is gradually gaining momentum in many developing countries to work out appropriate legislative systems that would be acceptable to the international community, and it is realized that the protection modalities have to be through a novel sui generis system. The foremost endorsement in securing such protection is offered in Article 8j of the CBD, providing that each CP shall respect, preserve and maintain knowledge, innovations and practices of indigenous and local communities and promote their wider application. Several hypotheses have been propounded, discussed and contested, generating new ideas and letting understand the complexities involved as regards the protection of TK. Some of the instruments advised to be used by most countries to protect the TK are discussed in the following sections.

Prior Art

The present IPR system, as recorded unambiguously in the Agreement on TRIPS, would render any knowledge non-patentable if it is a *prior art*, and so long as TK remains labelled as *prior art* because it is widespread among the indigenous and local communities, it cannot attract the protection of patent. The issue is being examined at various levels including at the WIPO whether TK can get such a protection, and it is asserted by many countries that TK should not be reckoned as *prior art*. The community or individual knowledge that has not been catalogued in publicly accessible databases should not be considered *prior art*. This connotes that TK should also be considered a patentable subject so long as it has not already been commercialized at certain level.

Deviating totally from the above is another school of thought that advocates proper documentation of all the TK that is held by the indigenous and local communities, basically aiming that that could be a safeguard against misappropriation of the TK, a practice endorsed by many pharmaceutical industries, making commercial use of the ethnobotanical information available as research results or published in some or the other form. The issue of documentation of TK is described in more detail later in this chapter.

Prior Informed Consent (PIC) Requirement

This arrangement of protection of TK requires that PIC shall be obtained in accordance with customary laws from the local communities for the use of their genetic resources and TK. However, practical difficulty of obtaining PIC represents an important constraint to benefit sharing with local communities. For foreign users or in-country scientists, the challenge is to know from which community PIC is required, and how to undertake the consultation without excessive complication or delay. PIC will be facilitated if indigenous and local communities are represented on a committee or by a focal point established to approve access applications (Swiderska, undated). Recently, the PIC has been made an indispensable part of many ABS regulations in a number of countries.

This arrangement is now well recognized, and according to the 'Guidelines for Research, Collections, Databases and Publications' drafted recently by the International Society for Ethnobiology, no research, collection, database or publication shall be undertaken without the PIC of all potentially affected communities of indigenous peoples or traditional societies (Dutfield, 1999).

Material Transfer Agreement (MTA) and the Certificate of Origin

This instrument requires that every patent applicant must declare that an invention being claimed is derived from the material and knowledge obtained lawfully and rightfully and also that due compensation has been paid to the providers of the resource. In the context of the international trade, bioprospectors entering into partnership agreements may have to provide a certificate from the national authority of the country and, if necessary, the donor agency that provided the resource, specifying that access was granted on the basis of PIC and on mutually agreed terms (MAT). In case the country providing the resource does not require an MTA, then a sworn statement to that effect has to be provided (Ghate et al., 1999).

While employing the condition for the certificate of origin, the habitats or the countries in question that harbour the biological source in natural or naturalized conditions have to be specified, and there are suggestions to give a cut-off date in the history (such as AD 1500) to determine prior natural geographical distribution of organisms (Ghate et al., 1999; Swiderska, undated). The idea of MTAs has received endorsement in many countries where the ABS regulations provide for such arrangements. In a report of an international workshop, held in 1998, submitted in the fourth meeting of COP to the CBD, among many concerns about the ABS, one was about the unapproved use of genetic resources in academia and ex situ collections (SCBD, 1998b). Incidentally, this report was submitted by the developed countries, and the response to this concern as suggested in this report was to encourage the use of adequate MTAs.

The issue of the certificate of origin is quite intricate as Carvalho (2003) contends that requiring the declaration of origin of materials or the PIC would be TRIPS-consistent only if, besides being reasonable for the purposes of Article 62 (Acquisition and maintenance of IPR and related *inter partes* procedures), the requirement extended to all fields of technology and to confine the requirement to the area of biotechnological inventions is an act of discrimination under Article 27.1. However, as McManis (2004) observes, Dr Carvalho argues quite persuasively that a conflict with the Agreement on TRIPS can be avoided if the disclosure of origin and PIC requirement is made a condition for enforcing the patent, drawing on established patent law doctrines of fraudulent procurement and unclean hands generally. McManis stresses that making the requirement a condition for enforcing a patent will serve to focus the inquiry into the adequacy of the disclosure of origin and evidence of PIC on those few issued patents that are sufficiently valuable to give their owner an economic motivation to bring infringement proceedings in a court of law.

One of the stated objectives for requiring disclosure of source of origin and PIC is to encourage compliance with the ABS principles of the CBD, and as an upshot, legal provisions are also being resorted to. For example, the Patents Act of India, as amended by the Patents (Second Amendment) Act, 2002, provides under Section 10 (Contents of specification) that the applicant must disclose the source and geographical origin of any biological material deposited in lieu of a description. Also Section 25 (Opposition to grant of patent) as amended allows for opposition to be filed on the ground that 'the complete specification does not disclose or wrongly mentions the source or geographical origin of biological material used for the invention'.

Sui Generis Legislation

Accepting the inadequacy of the existing IP system with regard to the protection of TK, a number of countries have enacted or are in the process of enacting sui generis system of protection of their genetic resources as well as the TK associated with these resources. The term sui generis, a Latin phrase, literally means 'of its own kind' or 'unique in its characteristics' and is widely used to refer to articles related to the legal discipline including IPRs. The term thus identifies a legal classification that exists independently of other categorizations because of its singularity and specific obligation. In the present context, Wekesa (2006) very aptly describes sui generis system of legislation for the protection of TK and the biological resources as:

> In IPR discourse, the term *sui generis* refers to a special form of protection regime outside the known (legal) framework, especially tailored to meet a certain need. In the African context, this regime becomes necessary in protecting traditional knowledge and associated natural resources.

> Traditional knowledge does not neatly lend itself to protection using the existing legal regimes because it is premised on the concept of community property ownership whereas the existing forms of IPR regimes are based on the Western concept of property ownership.

What makes an IP system a sui generis one is the modification of some of its features so as to properly accommodate the special characteristics of its subject matter, and the specific policy needs that led to the establishment of a distinct system (WIPO, 2002). The WTO Secretariat puts it more clearly by stating that the sui generis protection gives members more flexibility to adapt to particular circumstances arising from the technical characteristics of inventions in the field of plant varieties, such as novelty and disclosure. It is also believed that as sui generis arrangement is specially designed to achieve a certain purpose in a particular context, and no mandatory international rules or standards (which may be at odds with national objectives) exist, this implies that in order to develop an effective sui generis system of protection, its objectives should be clear and the goal should focus not only on the rights over the commercial use of TK but also to encompass the commercial, conservational and developmental goals (WHO, 2001a).

Now there is a continuing pressure for the establishment of an international sui generis system, as recently articulated by the G15 Group of developing countries, and the countries such as Bangladesh, and organizations such as the African Union began considering sui generis legislation that would provide community-based rights over biological resources and associated TK, seeking to give increased recognition to the cultural and customary practices of communities (CIPR, 2002). In many other tropical countries, the sui generis system of protection of TK is taking account of customary and local laws while these countries are progressively adopting effective sui generis regimes. For example, getting support from the Biological Diversity Act of 2002 along with the amended provisions of the Patents Act 1970, India now has an adequate system of protection of TK in place, and this illustrates how sui generis system is becoming a well-recognized legal framework for safeguarding the biological resources as well as the TK. Among other countries, for example, Brazil, Costa Rica, Peru, Portugal, Thailand, the African Union, the Philippines and Venezuela have all adopted sui generis measures that address TK and associated genetic resources

(WIPO, 2003b). The Philippines legislation on sui generis system takes the customary laws as a basis and the draft sui generis legislation of Namibia includes the mechanism of a community register. The efforts are on at documenting orally transmitted knowledge in Namibia and the Southern African region that are valuable primarily for the indigenous people. Whether these national systems as they evolve will have sufficient common characteristics to enable the development of an international sui generis system remains to be seen.

The above overview of the probable instruments for protection of TK gives rise to an understanding that with practical experiences at national or regional level, it should be feasible to develop an international system for the protection of TK. The CBD also has been pursuing the CPs to recognize, support and develop sui generis systems through national guidelines (SCBD, 2007). It may be true that no single agency has all the necessary expertise or resources to handle all aspects of TK. It may be inferred that a multiplicity of measures shall be necessary to protect, preserve and promote TK. When a wide range of biological resources and wider range of knowledge associated with the resources spread over diverse countries and communities are to be protected, it may be that a single all-encompassing sui generis system of protection may be too specific and not flexible enough to accommodate the more specific local needs.

The Aspect of Documentation of TK

It has been discursively put forth above how new concepts and novel ideas generated during the recent past for understanding the intricacies involved as regards the protection of TK. The foremost stress on such protection is probably offered in Article 8(j) (*Equitable sharing of the benefits of traditional knowledge*) of the CBD. From the member countries (CPs to the CBD), the ambit of Article 8(j) of the CBD expects that appropriate action should be undertaken to:

1. Preserve and maintain knowledge, innovations and practices of indigenous and local communities embodying traditional lifestyles relevant for the conservation and sustainable use of biological diversity—in short, the TK;

2. Promote wider application of TK with the approval and involvement of the TK-holders; and
3. Encourage the equitable sharing of the benefits arising from the utilization of TK.

The underlying tenets with respect to the documentation efforts of TK evenly comply with these provisions, and it is argued that the primary need for this purpose should be to document and preserve the TK that is held by the indigenous and local communities. The emphasis is on direct involvement of the indigenous and local communities, and surely no documentation of any TK should be undertaken without their consent, and, to be more specific, with 'their approval and involvement'. The motive is to preserve and maintain TK in order to ensure 'the conservation and sustainable use of biological diversity' that falls within the territory of the habitat of the respective communities. And lastly, there has to be an equitable sharing of the benefits that would accrue from the (commercial) use of the biological diversity and the TK associated with the biological diversity. If read between the lines, with the limitations that the Agreement on TRIPS puts on the scope of IPR protection to the TK, one would readily concur that the documentation of TK is the best alternative if the ultimate goal of an equitable sharing of the benefits has to be achieved. This approach should also be taking care of the apprehensions related to the misappropriation of TK, which is the greatest fear among the biodiversity-rich tropical countries.

Further arguments entailing the fundamental issue of benefit sharing begin from here and cover many options for the protection of TK. Though the preferences in general include sui generis system for protection of TK and the PIC, and lead to the debate on the issue of *prior art* and creation of databases and digitization, and so on, the very first apprehension that occurs as regards the documentation of TK is that it would be deemed to be in public domain—would fall into the category of *prior art*—once it is documented. Correa (2001) asserts that documentation of TK will not ensure benefit sharing with the TK-holders as it may even foreclose that possibility, to the extent that the documented knowledge is deemed part of the *prior art*. The argument put forth in favour of the documentation of TK is that once documented, though documentation would render it to be unpatentable, its likelihood of being misappropriated at least shall be lessened, and simultaneously, the sharing of benefits accruing from its commercial use in some way shall be feasible.

So the central issue in question remains the equity in benefit sharing. Referring to this issue, Nair (2002a) believes that while disclosure of the source of knowledge or the genetic resource in a patent application is an acknowledgement of their use, it does not provide for a reward system to the owner of that property. In fact, the procedures for actually examining patent applications have, in broad terms, not been designed to consider TK in the search of *prior art* even if, in principle, there is no reason why this should not be so. Ruiz (2002) observes that possible option that the TK is not misappropriated through the patent system could be ensured through the review of patent application examination procedure, taking into account all the accessible and available information contained in databases, publications and other sources of TK. The advantage with the documentation or the databases of TK is that these could be used in a practical and efficient manner by patent searching authorities. It is also recognized that there needs to be some degree of standardization if this mechanism is to be effective at the international level, though of course with right kind of planning and discussions with communities and other stakeholders, a consolidated set of tools may evolve which, in essence, will make TK much more easily and readily accessible throughout the world for a wide-ranging set of possible uses.

Dutfield (1999) argues that (a) not all traditional ecological knowledge is in the public domain (as the widely distributed and long-documented TK may be claimed to be so, it does not apply to more localized knowledge held by small numbers of people or an individual), and (b) unconsented placement of knowledge into the public domain does not in itself douse the legitimate entitlements of the holders and may in fact violate them. Dutfield further argues that unless the indigenous peoples have agreed to share their knowledge, documenting and/or disseminating their knowledge is morally wrong.

Thorpe (2002) observes in his study sponsored by the Commission on Intellectual Property Rights (CIPR) set up by the British government that of the countries studied, five countries (Egypt, Costa Rica, Bolivia, India and China), together with the members of the Andean Community, clearly required a patent applicant to disclose the source of biological material used in, or to develop, the invention. Thorpe adds that the Costa Rican legislation (Article 80 of Biodiversity Law of Costa Rica) on access to genetic resources also requires the applicant to present a certificate of access to acquire the genetic resources, and similar provision relating also to TK is provided in Decision 486 of the Andean Community. As a matter of fact, the Commission on IPRs endorses the need for protection of TK and genetic resources of developing countries with electronic libraries, and it also endorses the requirement of PIC and sharing the benefits of commercial exploitation (RIS, 2003).

There is no doubt about the fact that the patent rights offered by Article 27 (*Patentable Subject Matter*) of the Agreement on TRIPS are so enormous that all the legislative systems that are now taking shape in most of the tropical countries may finally prove to be poor and ineffective if an indigenous and local community seeks to earn benefits from the genetic resources when they are exploited commercially by a third party. There is a compulsion now on the biodiversity-rich developing countries such as India, Brazil and South Africa, and these countries are trying persistently and tediously at various forums, for incorporating suitable changes in the Agreement on TRIPS. The apprehension persists that the protection of the TK is still too vulnerable to be protected only through domestic laws of these countries, and it is contended that tough action would be required on the ground if TK was to be protected.

Alongside the anticipated suitable changes in the Agreement on TRIPS, the approach in this direction should be to encompass a few other interventions too. For example, the concept of 'public domain' should be redefined and technical issues such as the problem of collective or joint ownership and the modes of enforcement of rights should be resolved. Also, the registers of TK, for example, the *People's Biodiversity Registers* (PBRs) as included in India's Biological Diversity Act, or an easily navigable computerized database of documented TK (e.g., India's *Traditional Knowledge Digital Library* [TKDL] system) should be developed. For, once databases become widely available, this will enable patent examiners to conduct precise *prior art* searches in the field of TK.

With a view to prove the linkage between the habitat wherefrom a biological resource is extracted and the TK originates with the industrial product or outcome, the system of obtaining the certificates of origin could be introduced, obliging patent applicants to provide evidence of PIC for obtaining the biological resource and the associated TK from the country of origin and the indigenous and local community. Under this system, every patent office the world over should insist that the patent applicants declare that the knowledge and resources used in the patent applied for have been obtained lawfully and rightfully. It has to be understood genuinely that the IPRs are one of the tools being used, which have some limits and implications.

It would be pertinent to examine the complexities involved in the documentation aspect of the TK particularly with reference to the now quite prevalent modes of documentation, such as the PBRs, the TKDL and the Registration System. There are some good efforts made by individuals, research institutions, governments and NGOs that are included in the following paragraphs.

People's Biodiversity Registers

A proper documentation of TK may prove helpful in ensuring adequate protection, as otherwise it becomes impossible to defend misappropriation of the IP evolved and built up by indigenous people over a long period of time, often centuries, of intergenerational efforts. The documentation does not contribute to the development of an alternative to the patent system, yet it serves to show that the knowledge already exists and thus cannot be patented (Cullet, 1999). There are circumstances when documenting the TK may also prove counterproductive, if the IPR of the generators and holders of such knowledge are ignored by those doing the recording and if the archives are inaccessible to the communities providing the knowledge to the archives (Dutfield, 1999).

Several developed and developing countries have agreed on the importance of documenting TK. Firstly, this approach appears primarily flawed, as once published, novelty on the disclosed information cannot be claimed. However, there is another school of thought advocating maintenance of PBRs with some deviation from the basic concept of industrial patents. Just like the PBRs as provided for in India's Biological Diversity Act of 2002, the Peruvian Biodiversity Act of 2004 (the Act on the Protection of Access to Peruvian Biological Diversity and the Collective Knowledge of Indigenous Peoples—Law

No. 28216) also has provision of maintaining a 'register of biological resources and collective knowledge of indigenous peoples'. The WIPO also has a 'Traditional Knowledge Documentation Toolkit' project, which has been developed with an aim to help indigenous and local communities in identifying and defending their IP-related interests when their TK is documented or otherwise recorded.

Basically, the outcome anticipated from the documentation of TK in any form is that there is a record created that depicts that a particular idea or innovation has existed and there are people who are the first innovator of that idea.

Traditional Knowledge Digital Library

Following the patents granted on the products developed from the TK related to neem by the European Patent Office (EPO) and from the TK related to turmeric by the US Patent and Trademark Office (USPTO), a significant initiative was taken up by India to forestall grants of such patents in future. India initiated the project in 2001 to prepare an easily navigable computerized database of documented TK (which is already under public domain) relating to the use of medicinal and other aromatic plants, known as TKDL. Establishment of the TKDL was spearheaded by the Department of Indian Systems of Medicine and Homeopathy (ISMH). Its blueprint was prepared through the efforts of ISMH and three other government departments, the Council of Scientific and Industrial Research (CSIR), the Department of Industrial Policy and Promotion and the National Informatics Centre (NIC), and an interdisciplinary task force comprising experts from CSIR, Banaras Hindu University, NIC, Controller General of Patents and Trade Marks and Central Council of Research of Ayurveda and Siddha was set up to develop the proposal.

The TKDL compiles the information in digitized format in languages and format understandable by patent examiners in five languages (English, French, German, Japanese and Spanish). The project is being implemented at CSIR, and it engages the Traditional Knowledge Resource Classification (TKRC), a modern innovative structured classification system, based on the structure of International Patent Classification, for the purpose of systematic arrangement, dissemination and retrieval. Presently, the information contents for establishing *prior art* are available in 30 million A4-size pages, making the TKDL an effective tool for defensive protection of TK. The CSIR has concluded access agreements entailing institutional arrangements with the most important IP offices in the world including the EPO having 34 member states, the USPTO, Japan Patent Office, German Patent and Trademark Office, UK Intellectual Property Office, Canadian Intellectual Property Office and Intellectual Property Australia.[4]

On the basis of TKDL evidences, numerous patent applications have been withdrawn, refused or amended. India's TKDL is a unique tool in the sense that it enables prompt and inexpensive cancellation or withdrawal of patent applications relating to the country's TK. Still the misappropriation of the TK and biopiracy are of great concern in many countries today, and a global framework to protect the TK has yet to be established.

The Registration System

Arguing that it is difficult for a community to seek protection of its knowledge and inventive recipes, Gupta (1999a) suggests developing a registration system that will prevent any firm or individual to seek patent on community/individual knowledge and innovations without some kind of cross-licensing. Such a system would acknowledge the individual and collective creativity, and grant entitlements to grassroots innovators for sharing returns that may arise from commercial applications of their knowledge, innovations or practices with or without value addition. Professor Anil K. Gupta frequently refers to the Society for Research and Initiatives for Sustainable Technologies and Institutions (SRISTI) that made a proposal in 1993 for developing the International Network for Sustainable Technologies, Applications and Registration (INSTAR). The idea behind this system is detailed by Gupta as follows:

1. The register will help small investors seek opportunities of communication with communities and individual innovators

---

[4] *Source*: www.tkdl.res.in (Accessed in May 2015).

and explore opportunities of investment. A large number of potential negotiations will take place, increasing the opportunities for innovative communities and individuals.

2. Each entry in the Register will be coded according to a universal system so that geo-referencing of innovations can be done.

3. Later, some of the innovations will be considered appropriate for award of inventor's certificate or a kind of innovation patent, which is a limited purpose and limited duration protection.

4. The award of certificate will also increase entitlement of innovators for access to concessional credit and risk cover so that transition from collector or producer of herbs to developer and marketeer of value-added products can take place in cases where innovators deem that fit.

5. The registration system will also promote people-to-people learning and serve as a clearing house for indigenous and local communities. Wherever necessary and possible, formal scientific institutions will be linked up in the network.

With reference to a project in Ecuador entitled 'The Transformation of Traditional Knowledge into Trade Secrets' (SCBD, 1998a), it is argued that as a parallel to patents, copyrights and trademarks, which are accepted as instruments to enable the emergence of a market for information goods, oligopoly rights over genetic resources should be allowed to enable the emergence of a market for habitats. Thus, the project attempts to achieve a cartelization of TK within Ecuador. The project sets out to catalogue TK and maintains the database at regional centres, which is safeguarded through a hierarchy of access restrictions. The registration of TK and innovations can also form part of legislative systems designed for positive or defensive protection. Simultaneously, the documentation can also play a defensive role, as it can take many shapes and forms, through written registries and files, video, images and audio, in traditional indigenous language or others, and using modern or more classical technologies. The register forms a database that may well enable indigenous and local communities to keep a track on what happens to the TK they treasure, and Gupta (2001) recommends that these databases should be maintained in local language in order to ensure that such databases promote horizontal learning among people.

India's Biological Diversity Act of 2002 also provides for registration system to protect the knowledge of local people under Section 36(5) of the Act, which reads:

> The Central Government shall endeavour to respect and protect the knowledge of local people relating to biological diversity, as recommended by the National Biodiversity Authority through such measures, which may include registration of such knowledge at the local, State or national levels, and other measures for protection, including *sui generis* system.

The above stances establish that a register is not only a list or database designed to provide information to users—it is a database that is used to put information in order to gain legal rights relating to that information. A register of TK could serve to identify the existence of certain elements of TK over which rights exist and identify the persons, indigenous peoples or local communities holding those rights (WIPO, 2012a).

Out of the different forms or modes of documentation for having a compact database of TK, as described above, for the sake of protection of TK, the biodiversity-rich tropical countries are in the process of adopting one or the other and formulating sui generis system of registration of the TK belonging to the indigenous and local communities. A widely accepted fear about the TK being packaged into the database is that once the process of the creation of databases and digitization is complete, the issue of *prior art* will arise and the documented TK would be deemed to be in public domain. Then all the efforts will be of no avail as once the TK falls into the category of *prior art*, a situation will arise that will foreclose all the possibility of benefit sharing with the TK-holders. It has been argued above that to forestall the situation, the concept of 'public domain' shall have to be redefined. It would be pertinent to elaborate more on the issue of 'public domain' in this context.

The Imbroglio of Public Domain

It is commonplace that any knowledge that is in public domain is a *prior art* and cannot be patented. In the context of TK, therefore, it is almost conclusively stated that the concept of public domain needs to be redefined on the ground that TK is held either by an individual (traditional healer, etc.) or by small numbers of people and are confined within a small community that cannot be categorized as public domain. Dutfield (1999) argues that unless the indigenous peoples have agreed to share their knowledge, documenting and/or disseminating their knowledge is morally wrong. But then it is also argued that with respect to the community intellectual rights (CIRs; term coined by the Third World Network [TWN][5]), the purpose of preventing others from patenting will be achieved by publishing the local knowledge and making such publications available to the patent offices. Starting off from here, work has already begun in several cases, as discussed in the previous sections, for full-scale documentation of TK.

The initiatives for documentation include the PBRs prepared at the local level, documentation of the CIRs, the TKDL, INSTAR's registration system and so on. It is construed from the analysis of these initiatives that:

1. The relevance of international IPR regime to the CBD is beyond doubt (Dutfield, 2000), though the Agreement on TRIPS does not help at all in recognizing the contribution of TK.
2. It is well established that the application of TK and technologies can add value to genetic resources, and this ought to be a guiding principle in the bioprospecting partnership programmes.
3. While patents are clearly unsuitable mechanisms to protect the rights of TK-holders, the use of other IPRs may in some circumstances be feasible.

There have been examples during the recent past of misappropriation of TK-based innovations that are an indication of their vulnerability. The limitations in the protection of TK were identified in a previous

---

[5] TWN is an independent nonprofit international network of organizations and individuals involved in issues relating to development, developing countries and North–South affairs. Formed in 1984 in Malaysia especially to strengthen cooperation among development and environment groups in the South, its mission is to bring about a greater articulation of the needs and rights of peoples in the South, a fair distribution of world resources and forms of development that are ecologically sustainable and fulfil human needs. *Source*: www.twnside.org.sg (Accessed in November 2014).

section wherein the reasons why the existing IPR regime is supposed to be inappropriate for the protection of TK have been put across. What is of concern is that nothing has yet been determined as final regarding the legal protection of TK. To some extent, the national laws that have come up or are in the process of being evolved in many developing countries seem to be the mechanism, and WIPO also, among many other institutions, is undertaking extensive work on the aspects pertaining to the protection of TK and the ABS arrangements.

There is now an almost general agreement that proper documentation of TK could help ensure adequate protection, and without that one would find that misappropriation of the TK held by the indigenous people is too difficult to defend. It is readily agreed that the IPR regime dealt in the Agreement on TRIPS and other instruments was not evolved to deal with the TK, and once published, novelty on the disclosed information cannot be claimed. However, the stand that every patent office in the technology-rich 'North' ought to insist the patent applicants to declare that the knowledge and resources used in a patent have been obtained lawfully and rightfully seems valid in this respect and this would call for an approach for documentation of the TK system.

It is consequently advised that the intention of preventing others from patenting ought to be achieved by publishing the TK and practices and making these publications available to the patent offices the world over. The amount of available information is substantive, though it is not in a form that would make it easily navigable and searchable by patent examiners. As a result, efforts have already begun for an easily navigable computerized database of documented TK relating to the use of genetic resources. The steps include trying the instruments such as the PIC requirement, the MTAs and the certificate of origin. A significant support to these efforts is offered under Article 8(j) (*Equitable sharing of the benefits of traditional knowledge*) of the CBD, providing that each CP shall respect, preserve and maintain knowledge, innovations and practices of indigenous and local communities and promote their wider application.

Lastly, we realize the importance of TK associated with the biological resources held by the members of the indigenous and local communities, and find that the present IPR regime is insufficient to protect the TK against misappropriation. In fact, an acceptable solution for the protection and promotion of TK has yet to emerge despite the constant efforts made under the new IPR regime. A large number of bodies including the WIPO are engaged in the debates and discussions for the protection of TK, and yet no universally acceptable mechanism has been developed so far. In the meantime, many tropical countries either have formulated or are in the process of formulating laws for sharing of benefits accruing from the commercial use of genetic resources and the TK associated with them. An issue of major concern is that the TK is bearing the risk of disappearance in future, and the indigenous and local communities are helpless if there is any claim over the benefits that are drawn when their TK is made use of commercially, merely on the ground that the information is now in the public domain. It is also stated that the concept of public domain needs to be redefined on the ground that TK is held either by an individual or by small numbers of people and are confined within a small community that cannot be categorized as public domain. The ambiguity still remains, and multitudinous options and alternatives are suggested for the protection of TK against its misappropriation and for sharing of benefits drawn from its use, such as: arrangement for PIC to be obtained from the local communities, the instrument of MTA and the certificate of origin that requires that every patent applicant must declare that an invention being claimed is derived from the material and knowledge obtained lawfully and rightfully and also that due compensation has been paid to the providers of the resource and the instruments such as PBRs, the TKDL and the Registration System that involve documentation of TK. A broadly accepted suggestion is that the present IPR regime has to be reformed in order to protect the TK, and the CBD also has been pursuing the CP to the CBD to recognize, support and develop sui generis systems through national guidelines for the protection of TK.

The globalization set off with the beginning of the WTO and its new-found IPR regime leave no option for the developing countries except to go for developing sui generis system, unless the desired reconciliation between the Agreement on TRIPS and the CBD is arrived at. It is well understood now that under the sui generis system and the creation of database for documentation of TK as per the local laws may ascertain to limit the misappropriation of TK. This would also be of help in looking for the prospects of formulating mechanism to draw monetary gain from the commercial use of these knowledge systems.

4

Implications of WTO and the North–South Divide

It is well accepted that out of the 60-odd agreements and decisions that are part and parcel of the agreement establishing the WTO, the Agreement on TRIPS is most crucial as regards the commercial use of the biological diversity of tropical countries with its maximum impact on two fields of the industry: the pharmaceutical and the agricultural. In the present chapter, an attempt is made firstly to investigate the anomalies in benefit sharing, followed by scrutiny of the various Articles of the Agreement on TRIPS and the CBD, an assessment of the limitations of these two international treaties and the North–South divide often perceived in this context, and in the end, it has been evaluated as to how far the CBD may be helpful in circumventing the implications of the Agreement on TRIPS. In the analysis, though the WTO in general is labelled as an instrument working against the interests of the developing countries and what is often talked about is the 'implications' only, the positive aspects also are covered against the backdrop of the negative pursuits perceived as working against the objectives of the CBD and against the interests of the developing countries.

A detailed account has been given in the previous chapter of the various modes for protection of the TK held by the indigenous and local communities under the realm of the present IPR regime in order to avert the misappropriation of benefits drawn from them. However, the sharing of benefits from the TK associated with these resources is not straightforward, and the arrangements for the sharing of benefits accruing from the genetic resources among various stakeholders are highly skewed in most of the partnership programmes today, leading to serious anomalies occurring in the process. An attempt for proper understanding of these anomalies has been made in the following section.

Anomalies in Benefit Sharing Arrangements

The roots of the anomalies, as mentioned in the previous chapter, lie probably in the most innocuous act of ethnobotanists who extract the local knowledge, publish it without any attribution, reciprocity or benefit sharing (Gupta, 2001), and this knowledge ultimately enters the global industrial market that has developed considerably during the recent decades in the field of pharmaceuticals and other nature-based products. Gupta (2001) points out that the rate of erosion of local knowledge about biodiversity has never been as high as it is in the current generation. This happens because the ethnobotanical knowledge is usually obtained from published sources rather than directly from indigenous and local communities (Swiderska, undated).

There are several reasons for these anomalies. Firstly, the scheme of equitable sharing of benefits is not supported by the current institutional arrangements, particularly the existing system of IPR. Secondly, the biological resources are gathered typically from the established sources such as the gene banks and botanic gardens that have flourished during the recent decades in the industrialized world (Swiderska, undated). Referring to this asymmetry in benefit sharing, Gupta (2001) further states that the pressure on natural resources is increasing, but the livelihood options of the local communities and disadvantaged people in many cases have not increased. Under such a situation, the concerns about asymmetry in the sharing of benefits through the use of TK and biodiversity resource are bound to become more acute and urgent. The charges of biopiracy against various transnational corporations and research organizations continue to be levelled by the NGOs and other civil society actors because of lack of reciprocity in sharing of benefits.

Walter (2002) also indicates towards this anomaly, stating that the increment in the market value may not be equally shared among all the stakeholders involved at different levels, while the benefits from the commercial use of these genetic resources are largely being enjoyed by companies and research institutes in the 'North'. It is now well evidenced and ascertained that the private sector manufacturing organizations have the technology for product development, and can obtain IPRs and patents on novel products to protect investments in R&D. Giving a profile of the innovations and breeding by farmers and peasants, Vandana Shiva (2000) describes that for centuries the Third-World farmers have evolved crops and given us the diversity of plants, and this innovation by farmers has not stopped. She observes that centuries of collective innovation by farmers and peasants are being hijacked as corporations claim IPR on these and other seeds and plants.

The issue of sharing of benefits from these collective innovations is largely debated alongside the discussions on biodiversity conservation at various platforms, especially in the context of the international trade. The main benefit of the international trade in genetic resources is the high market value the products achieve compared to the value the products fetch in local or national markets, chiefly after well-matched value addition and the new trends that exhibit a fast-growing demand for nature-based products. However, high market values combined with high demands may also cause unsustainable use since they may lead to the overexploitation of the favoured species. Additionally, as Walter (2002) mentions, the higher product values may not be equally shared among all stakeholders involved in the collection, processing, manufacturing, trade and marketing of the products of biodiversity. It is also observed that almost all pharmaceutical companies use some in situ material collected from forest areas or community lands, and this demand is likely to continue, unless national access legislations become restrictive.

The matter of access to the genetic resources related to these innovations sometimes involves a basic question: Why should the benefits be shared at all? Numerous arguments have been listed by authors why entrepreneurs must share the benefits accruing from the genetic resources and the associated TK with the indigenous communities who hold such knowledge. It is maintained that the indigenous people have intricately been dependent on these resources for centuries for their livelihood and health-care needs, have the rights to draw usufructs from these resources, and they should not be denied access unless they are provided with workable alternatives. Gupta (1999a) frames his arguments on the fact that given the high hit-rate in the research based on identified uses of biodiversity, costs of formal R&D systems are considerably reduced. He asserts that local communities should be considered co-inventors of the new value-added products, and that TK should not be treated as common property. Referring to the present trend of benefit sharing arrangements, Swiderska (undated) mentions that although benefit sharing with local communities is still uncommon, a number of examples exist, particularly in the pharmaceutical and herbal medicine sectors, for some kind of sharing arrangements.

A new trend in benefit sharing is now almost established, particularly after the CBD became effective. Moreover, the principle of benefit sharing is now not confined to the tenets of the CBD only, and besides having a number of examples of being in practice, a more pervasive document, the Swiss Guidelines on ABS (Girsberger et al., 1999) also mentions about it:

> Subject to international law and national legislation, the holders of intellectual property rights based on genetic resources are encouraged to share their intellectual property rights with other stakeholders who contributed to the conservation of these genetic resources or to the scientific research and development based on these genetic resources. (Article 8.3)

Unfortunately, complications are frequent in the ABS arrangements of the present day. The broadening of IPR regime set in motion the world over by the Agreement on TRIPS has raised new concerns and problems, and it does not seem to have yielded much benefit to the developing countries. The benefit sharing arrangements call for more deliberations at the international level, and what is needed is a well-organized approach to ensure greater transfer of technology to developing countries, which may involve not only certain changes in the provisions under the Agreement on TRIPS but also a possible review of other WTO agreements.

The issue of the transfer of technology itself has drawn heated debates after signing of the CBD. The USA was the only major country not to sign the CBD at Rio de Janeiro, and the argument was that the USA could not make commitments for private industry to transfer protected technology (Miller, 1995). In this context, it is rightfully remarked by Ramanna (2002) that evolving strategies to gain access to technologies that are protected under IPR laws would be important in the new patent scenario, as there are several options for accessing patented technology. Some options could be explored to effectively utilize one's own IP such as cross-licensing, patent pooling (an aggregation of IPRs that are cross-licensed to third parties) and clearing house mechanisms (bundling together sets of complimentary patents from different patent holders and creating customized licenses). Ramanna further states that strategies could be developed to promote sharing of technology—it has been possible to persuade some transnational corporations to provide cost-free licensing of technologies, and independent developers of research tools could provide some scope for collaboration for gaining access to technology.

In a nutshell, it is ascertained that a systematic approach is needed to ensure the transfer of technology from the 'North' to the 'South'. It is also ascertained that the pharmaceutical companies use some in situ material collected from forest areas or community lands, and that they must share these benefits with the indigenous and local communities who so far have not been able to draw benefits from bioprospecting. It is also argued that the demand for genetic material from the 'South' is likely to continue. What are the possible legislations that can ensure these propositions? Is it possible to earn returns from the ill-exploited riches of the TK, so far treated as a common property, in order to benefit the indigenous and local communities? Is it feasible to evolve a mechanism for the local communities to claim to be the co-inventors of new value-added products? High market values combined with high demands are causing unsustainable use through overexploitation of the species providing nature-based products—are adequate efforts being made for conservation of these species? Could an equitable sharing of benefits be ensured through some strategy that is able to remove the anomalies described above and ensure sustainable exploitation?

These and other related questions still remain unanswered and need detailed investigation in the light of the provisions made under the Agreement on TRIPS. The issue is taken up in the following section. Some of the benefit sharing programmes, spanning over the last 20–25 years, are examined in Chapter 6 to highlight the anomalies in benefit sharing as well as the unique provisions the programmes consist of that would move these programmes nearest to the equitable benefit sharing phenomenon.

The Advent of WTO

The WTO came into existence in 1995 as the multilateral trading system originally set up under the GATT. Interestingly, before 1995, for most of the history of the international trade regime, the GATT—the predecessor of WTO—attracted little attention, for it dealt almost exclusively with manufactured goods, and what happened at the borders when these goods were traded (Halle and Borregaard, 2004). The development and environment policies were left in the realm of domestic decision-making, or to purpose-built regimes dealing with their international aspects. With 161 members, the WTO acts as a forum for trade negotiations by administering trade agreements, settling trade disputes and reviewing trade policies and assisting developing countries in trade policy issues through technical assistance and training programmes.

It would be appropriate to give a brief outline of the WTO framework before reviewing the implications it has on the commercialization aspect of the components of biodiversity, particularly in the context of the international trade and the TK associated with genetic resources.

The WTO has a legal personality and its topmost decision-making body is the Ministerial Conference, held at least once every two years. The next in command is the General Council, entrusted with carrying out the functions of the WTO. Next levels are the Goods Council, Services Council and Intellectual Property (TRIPS) Council or the Council for TRIPS. The WTO Secretariat, based in Geneva (with no branch offices elsewhere), does not have any decision-making role since the decisions are taken by the members themselves. The WTO is headed by a director general and the strength of staff of the Secretariat of WTO is about 640. The staffing pattern in the WTO by nationality shows that 63 per cent of them hail from eight developed countries

(France, UK, Spain, Switzerland, USA, Canada, Germany and Italy; maximum 175 from France and minimum 16 from Italy), 10 per cent from other developed countries and only 27 per cent of them belong to the developing or least developed countries.[1] It is worth mentioning here that the majority of WTO members are developing countries (WTO, 2001).

The IPR issues were included in 1994 at the end of the Uruguay Round of trade negotiations under GATT as a result of an intense lobbying by the developed countries, specifically the USA, EU and Japan, and the Agreement on TRIPS formally became an instrument of the WTO. The Agreement on TRIPS, included as Annexure 1C of the Agreement Establishing the WTO (or the WTO Agreement, which serves as an umbrella agreement) is one of around 60 agreements and decisions that total about 550 pages. The ratification of the Agreement on TRIPS is a compulsory requirement of WTO membership for any country that wishes benefit from and enter into the numerous international markets opened by the WTO.

Remarkably, the Agreement on TRIPS is the main instrument in the globalization of patent system as it has widened the scope, duration and strength of patent protection. The impact of the Agreement in two fields, namely, agriculture and pharmaceuticals, has raised maximum controversy in developing countries. Ramanna (2002) argues that India initially resisted the inclusion of IPRs in the WTO but ultimately signed the Agreement, for the WTO was a 'take it or leave it' agreement—either a member accepts all the agreements or none, leaving no scope for partial agreement. That is how the new international trade regime went well beyond the area of international trade, and the new issues such as IPR, agriculture, investments and services were brought on the agenda of multilateral trade negotiations, leading to the most complicated rules formulated by powerful players (Chandiramani, 2002). After 1995, the type of patent system a country has to adopt, the agricultural support to be practised and the type of technology or pattern of investment followed, all these were to be decided as per the tenets of the WTO.

The limitations do not end there, as Halle and Borregaard (2004) state that the creation of the WTO represents an expansion of the scope

---

[1] *Source*: www.wto.org (Accessed in November 2014).

of trade policy, providing a wide range of fields—food safety, product standards, standardization, IPRs, services and environment—that were, before that, the exclusive preserve of national governments, or of specific intergovernmental regimes. They indicate that the power of dispute settlement system of the WTO, with the ability to impose sanctions and to ensure that the trade policy took precedence over other policies, does not remain only a matter of academic interest.

Though seemingly rational, the system has an innate inconsistency as the recent studies of the trends in the global trade show a clear lopsidedness that disfavours the poor in general. As per some citations (Mehta, 2002; Miller, 1995), 500 corporations control 70 per cent of world trade, 80 per cent of foreign investments and 30 per cent of the world's GDP, and top 10 companies control 40 per cent of the world's prescription drug market while top 20 companies held about 5 per cent in the early 1980s. These numbers unquestionably guide to the theory that these corporations shall be able to influence the policy making at the international level, prompting the resource-poor and technology-rich side to attain mega-influences. The benefit sharing programmes of the recent years that are discussed in Chapter 6 explicitly reflect these mega-influences.

In spite of these influences, it is well recognized that the national economies are now becoming more interdependent, guiding the process of globalization of trade and development. Carley and Christie (2000) observe that the globalization is a multi-stranded phenomenon: it encompasses cultural and political trends as well as economic processes that together form a dynamic complex of issues. This process, which brings more and more countries into trading networks as producers and consumers of goods on an industrial scale, leaves no option for any nation state of insulating itself from changes in the economic climate and from the policies of international actors such as the transnational corporations. Carley and Christie (2000) emphasize that as the transnational corporations develop, there is an increasing flow of managerial and specialist personnel across frontiers, and the development of computerized communication network has created global flow of information and capital that ignore frontiers, making it virtually impossible for developing countries to resist pressures to follow Western development paths. The observation is worthy of appreciation and is reflected convincingly with the ever spreading globalization in today's world.

An altogether different kind of remark articulated by Patnaik (1997) is that the international trade regime, highly skewed in the beginning in favour of the developed countries while maintaining some derogation in favour of the developing countries, is slowly emerging into a more balanced regime. In the context of the WTO vis-à-vis the fate of developing countries, Patnaik states that the rich countries have an enormous incentive to trade with the developing countries, and with the introduction of market economies and liberalization of economies of many developing countries, the prospects for an increase in the volume of trade in the twenty-first century are enormous. The remark is seemingly true as it is witnessed how the developing countries such as India, Brazil and China have restructured their economies with markets that are affecting foreign investment and trade. This new phenomenon may enable a more equitable sharing of benefits accruing from the richness of biodiversity of the developing countries, once their domestic trade regimes get assimilated into the world market. However, though such a scenario of optimism is not yet visible, it should be appreciated that international negotiations of such nature would carry on for long years probably because of the clash of business interests and the lack of mutual trust.

The Agreement on TRIPS

As has been stated earlier, the impact of the Agreement on TRIPS in two fields, namely, agriculture and pharmaceuticals (the root cause being its provisions related to patents), has raised the most controversy in the developing countries. However, in addition to the subject of patents, the Agreement on TRIPS covers a wide range of other IPRs, for example, copyright, trademarks, geographical indications, industrial designs and trade secrets, and a choice set of these instruments opens the scope of divesting the biodiversity-rich 'South' from its intrinsic apprehension that its genetic resources and the TK associated with them remain unprotected under the new patent regime.

It is relevant to be aware here that the Agreement on TRIPS is a framework agreement; it is to be operationalized via countries' national laws (WHO, 2002b). It is also important in this context that the Agreement on TRIPS is silent on the issue of sharing of benefits with indigenous and local communities. It is also frequently reiterated that for making sharing of benefits with local communities feasible, it would be necessary for IPR laws to have stringent norms of disclosure of the country and the community from which a patentable subject matter and information regarding its use was obtained, as well as proof of consent of the country of origin. Providentially, this observation is now seeing the light of the day in the newer ABS laws being formulated by many developing countries.

In a further scrutiny of the provisions under the Agreement on TRIPS, Nair (2002a) states that Article 7 (*Objectives*) defines the main objectives and principles, which mentions that IPRs should contribute to social and economic welfare and to a balance of rights and obligations of the members, while Article 8 (*Principles*) stipulates prevention of abuse of IPR that will restrain trade or adversely affect international transfer of technology (see Box 4.1). Nair observes that a liberal and permissible interpretation of these Articles will ensure equity in the application of the Agreement on TRIPS to the benefit of the developing country's economy and social welfare. But it is only in Article 7 of the Agreement on TRIPS that international minimum standards for the protection and enforcement of IPRs in general are prescribed (McManis, 2004). Halle and Borregaard (2004) contend that the 'first crack' in the TRIPS edifice was put by the Doha Ministerial Conference of 2001 by forcing a relaxation of IP protection for public health purposes. The Doha agenda now calls for framing future action on other relevant provisions of the Agreement on TRIPS, and an approach is needed to agree on a clear outline in the context of the WTO and IPRs, beginning with recognition of measures to protect the TK held by the indigenous and local communities.

In this regard, a veritable option could be that Article 29.1 of the Agreement on TRIPS is amended to include these requirements, especially the disclosure of the origin of the genetic resources and TK, as part of the patent application procedure. It can only be said that the battle is on as it is a real tough task to win on any economic front at the global level, and neither the developed countries nor the developing countries would so readily relent on this front. In the context of several other intricacies involved, more reflections on this issue are included in Chapters 8 and 9 of this book.

Box 4.1:
The Agreement on TRIPS: Article 7 (Objectives) and Article 8 (Principles)

The Agreement on TRIPS: Article 7

(Objectives)

The protection and enforcement of intellectual property rights should contribute to the promotion of technological innovation and to the transfer and dissemination of technology, to the mutual advantage of producers and users of technological knowledge and in a manner conducive to social and economic welfare, and to a balance of rights and obligations.

The Agreement on TRIPS: Article 8

(Principles)

1. Members may, in formulating or amending their laws and regulations, adopt measures necessary to protect public health and nutrition, and to promote the public interest in sectors of vital importance to their socio-economic and technological development, provided that such measures are consistent with the provisions of this Agreement.
2. Appropriate measures, provided that they are consistent with the provisions of this Agreement, may be needed to prevent the abuse of intellectual property rights by right holders or the resort to practices which unreasonably restrain trade or adversely affect the international transfer of technology.

Source: www.wto.org/english/docs_e/legal_e/27-trips.pdf (Agreement on TRIPS).

Patenting of Varieties of Genetic Material and Life Forms

Besides the aspect of disclosure of the origin of genetic resources and TK, another important feature related to commercial use of biodiversity is the prohibition of patenting of life forms. Under the provisions of the Agreement on TRIPS, biotechnology products and processes are protected in the broadest possible sense, though the Agreement excludes plants and animals other than microorganisms from patenting. In the WIPO round table meeting held in November 1999 to discuss how to bring the informal domain of TK into the formal IP system, Victoria Tauli-Corpuz, who is the Executive Director, Indigenous Peoples' International Centre for Policy Research and Education (Tebtebba Foundation), Philippines, reiterated the call of indigenous peoples all over the world against patenting of life-forms and life-creating processes, referring to the statement of over a hundred indigenous people's groups opposing patenting of life in the Agreement on TRIPS, which she said was consistent with several proposals put forward by developing countries (SUNS, 1999).

This issue is raised in almost every session of the WIPO's Intergovernmental Committee on Intellectual Property and Genetic Resources, Traditional Knowledge and Folklore (IGC) and in other international conventions and conferences, and there is constant endeavour to reach to some accord. Many remarks, papers and comments see the light of the day by numerous authors belonging to the scientific community, academic fields, NGOs and other groups to comprehend this issue, a few of which may be recounted here. For example, with reference to the provision under Article 27.3(b) of the Agreement on TRIPS, Rao (2001) observes that powerful voices are heard that by sanctioning the patenting of varieties of generic material developed over generations by the countries of the 'South', and enabling Northern corporations to secure monopoly over them, the Agreement on TRIPS is undermining the concept of equitable benefit sharing.

In the same context, Watal (2001) records that at the time of the negotiations, the USA and the EU differed on their approaches to patenting biotechnological inventions—the USA believed that anything under the sun made by man, except for human beings, was patentable, while the EU was grappling with strong internal resistance to patents on living organisms. This aspect needs close examination, for the developing countries also are opposed to this concept, though the motive differs. Their motive behind the disapproval is that the Agreement on TRIPS is aiding the exploitation of biodiversity by privatizing biodiversity expressed in life forms and knowledge.

The issue of patenting of the indigenous genetic resources and particularly life forms was debated at length in the fifth session of the IGC of WIPO held in 2003 (WIPO, 2003a), in which the French delegation, recalling that members of the EU had made a proposal regarding disclosure of origin of TK and genetic resources used in inventions, provided this did not become a criterion of patentability, pronounced that a system for positive protection could be envisaged and close study was needed of questions linked to patentability of life forms, and countries' choices needed a balance between economic and ethical considerations. In the same session, the representative of Genetic Resources Action International (GRAIN), speaking on behalf of the indigenous peoples of Colombia, expressed general concern with the discussion on TK and associated genetic resources and stated that the indigenous peoples had conserved genetic resources for thousands of years and rights to them were collective. Expressing opposition to genetic mapping and patents on life forms, the representative also stated that the type of protection that was discussed in the session formed part of a capitalist approach and regarded TK and associated genetic resources merely as ex situ inputs to markets, which was not in accordance with the perspectives of indigenous peoples.

The Cuban delegation in the same session stated that the clarification of measures and elements that limit acquisition of IPR for those who do not have any rights in the genetic resources was the proper way of understanding the term 'protection of genetic resources'. It stated that positive protection of forms of life and the viewing of genetic resources as a whole as well as its replication were a matter being dealt with by WIPO as a substantive patent issue.

Referring to the observation of the WIPO Secretariat that documentation for the purposes of defensive protection may actually facilitate the misappropriation of TK and the components of biodiversity, the Indian delegation in the same session stated that in majority of the cases, misappropriation of TK occurred due to the lack of access to public domain TK to the patent examiners that may be for the reasons of the non-availability and/or format and language barriers. The delegation stated that it was for this reason India had to fight for the revocation of patents such as turmeric and basmati at USPTO and neem at EPO by incurring considerable efforts and resources, and that was done only to establish the demonstrative effect on the issue of misappropriation of India's TK. It was suggested by India that as a remedy, a mechanism along with detailed guidelines should be formed for setting up of regional and international registries in the area of TK and the components of biodiversity.

The approach that India and other developing countries adopt to press upon this obligation has its origin in the CBD, the first treaty that talks about and acknowledges the vital role of TK, the customary use of biological resources in accordance with traditional cultural practices, the innovations and practices for biodiversity conservation and the risks associated with the living modified organisms resulting from biotechnology. Before deliberating on the approach being adopted by developing countries at the international forum to nullify the implications of the Agreement on TRIPS, a brief account of the CBD is given in the following section.

Convention on Biological Diversity

In 1972, the UN Conference on the Human Environment at Stockholm resolved to establish the UNEP. Governments signed a number of regional and international agreements to tackle specific issues, such as protecting wetlands and regulating the international trade in endangered species. In 1987, the World Commission on Environment and Development (the Brundtland Commission) concluded that economic development must become less ecologically destructive. In 1992, the largest ever meeting of world leaders took place at the UN Conference on Environment and Development (UNCED; informally known as the Earth Summit) in Rio de Janeiro, Brazil. An historic set of five agreements (Agenda 21, the Rio Declaration on Environment and Development, the Statement of Forest Principles, UN Framework Convention on Climate Change and UN CBD) was signed at the Earth Summit, the most debated agreement out of this set since the happening of the Earth Summit has been the CBD. This treaty is proclaimed to be the first global agreement on the conservation and sustainable use of biological diversity. The CBD establishes three major goals:

1. Conservation of biological diversity,
2. Sustainable use of its components, and
3. Fair and equitable sharing of the benefits from the use of genetic resources.

The CBD's ultimate authority is the COP, consisting of all governments and regional economic integration organizations that have ratified the treaty. The COP can rely on expertise and support from several other bodies that are established by the CBD. Since its adoption, the CBD has strived to implement its three major goals. The applicability of these provisions is brought in by way of the recognition under the CBD that biological diversity is a global asset, its conservation is a common concern of the humankind, and also that States have sovereign rights over their own biological resources. It appreciates that States are responsible for conserving their biological diversity and for using their biological resources in a sustainable manner, and that special provision is required to meet the needs of developing countries, including the provision of new and additional financial resources and appropriate access to relevant technologies.

As regards the access to genetic resources, the CBD requires that access wherever granted should be on mutually agreed terms and it also requires prior informed consent of the CPs while accessing biodiversity (Article 15). And probably one very important provision (Article 8j) states that the knowledge, innovations and practices of indigenous and local communities relevant to biodiversity conservation and utilization should be respected and be preserved and maintained by the CPs. It requires nations to protect the TK and customary practices relating to the uses of biological resources (Article 10c). It further obliges CPs to promote the wider application of such TK with the approval and involvement of the holders of TK and to encourage equitable sharing of the benefits arising from the utilization of the knowledge. Articles 15.6, 15.7, 16, 19.1 and 19.2 advocate for fair and equitable benefit sharing arrangements between the providers and the users of relevant resources.

The developing countries of the biodiversity-rich 'South' have frequently been looking to these provisions of the CBD to derive a broad framework for taking up the bioprospecting activities based on the wealth of their biological diversity and the associated TK. Of particular interest is the stipulation that the conservation of biological diversity depends not only on the sustainable use of these resources but also on the equitable sharing of benefits that result from their commercial use. Nevertheless, the CBD in effect is understood as tentative in nature, and it is believed that it needs support from other treaties or covenants if its provisions are to be made effective. The States also have to formulate suitable national laws in order to meet the CBD objectives. Some of the limitations are summarized in the following section.

Limitations of the CBD

The foremost objective of the CBD revolves around the conservation of biodiversity of the world, and the fact is that the tropical forests situated habitually in the Third World are the repositories of most of the biological resources, and most of the forested regions are cohabited by the indigenous people who traditionally have been dependent for their sustenance and health care on these resources in a common perception of mutual coexistence. Now these very resources suddenly have become a global asset that are to be conserved, used sustainably, and the benefits have to be shared equitably between the indigenous people and also among other stakeholders. Indeed, unless the benefits generated are equitably shared among the different stakeholders, source countries will find little incentive to conserve their biological diversity (Guerin-McManus et al., 1998).

Another striking observation on the part of the developing countries is that the CBD is the first international treaty to acknowledge the vital role of TK, innovations and practices in biodiversity conservation and the need to guarantee their protection, whether through IPR regime or other means. It is believed that if the TK and indigenous cultures and innovations are not capable of being protected by the existing IPR system, there is at the very least a moral obligation for governments to safeguard these entitlements either through a new IPR law or by other legal means.

In the above context, post-1995, the year when the WTO came into being, a trend is seen especially among the academicians and NGOs in the developing countries to look optimistically to the CBD. However, the CBD is silent about the ownership or property rights of the genetic resources, chiefly those that are already in the national and international gene banks (Sharma, 1996). Also, though the CBD highlights the need for benefit sharing with indigenous and local communities, it leaves benefit sharing policy to be defined locally, and correspondingly, it does not provide clear legal rights to the indigenous communities over their knowledge or genetic resources. Benefit sharing with local communities is at times limited because genetic resources are usually acquired from ex situ sources (gene banks and botanic gardens), and the majority of repositories of germplasm and other pertinent information are not bound by the CBD because they were collected before its entry into force and are now available through repositories housed in the developed countries (Ghate et al., 1999; Sharma, 1996; Swiderska, undated).

As such, the options are very limited to arrive at an equitable sharing of benefits accruing from the natural resources belonging to the biodiversity-rich countries. Jermy (1995) rightly believes that no legal claim can be made by a CP for the past—*and future*—use of any genetic resource obtained before the CBD came into force. Jermy gives an example that will explain how vulnerable the biological wealth of the world is: At Royal Botanic Gardens, Kew, in the last 10 years, phytochemical and biological studies have been carried out on over 4,050 species of plants out of the 25,000–30,000 species available in the living collection in the gardens. Another example is the basmati lines developed by RiceTec, a US company, from the strains obtained from India before the CBD came into force (Ghate et al., 1999).

Similarly, ethnobotanical knowledge is usually obtained from published sources rather than directly from indigenous and local communities, and many pharmaceutical companies use in situ material collected from protected areas or community lands, and this demand continues (Swiderska, undated). Commenting on the provision under Article 16 (*Access to and Transfer of Technology*) of the CBD, Clive Jermy (1995) quotes Glowka et al. (1994) to support that it is possibly the most complex and controversial article in the CBD.

Taking this discussion further, McLaughlin (2003) may be quoted here, observing that the CBD does not provide for compulsory and binding dispute settlement, and should a nation seek to enforce any rights or obligations granted under the CBD, it can only do so through nonbinding conciliation, further minimizing any assertion that the CBD poses a serious risk to IPRs. McLaughlin believes that despite the early US protest that the CBD does not protect IPR, in reality, it recognizes IPR to the full extent of existing international law. Ghate et al. (1999) concur with McLaughlin with the hope that it may be possible to safeguard the interests of the Third World taking help from many provisions of the CBD itself. The scrutiny of the arrangements of benefit sharing that were examined as case studies, detailed in Chapter 6, is based closely on these tenets contained in the CBD as well as the interrelated ideas and opinions expressed by several authors in this context.

The CBD is silent also about the rights of the indigenous and local communities, and it sets aside benefit sharing arrangements to be provided in the national laws. An important factor in benefit sharing with the indigenous and local communities is that majority of repositories of germplasm are not bound by the CBD because they were collected before the CBD came into being (Ghate et al., 1999; Sharma, 1996). The result is that no legal claim can be made by a CP of the CBD for the past or future use of any genetic material or knowledge transferred abroad prior to 1993, the year when the CBD came into force (Ghate et al., 1999; Jermy, 1995). Sharma (1996) also observes that the CBD is silent about the ownership or property rights of the genetic resources that are already in the national and international gene banks, and it specifically leaves the door open for the patenting of genetic material ostensibly in the interest of the developed countries, both by the private and by the public sectors. Giving the example of database of NAPRALERT (Natural Products Alert) at Chicago, an extensive source of information on traditional uses of Indian plants, Ghate et al. (1999) state that the information is compiled through exhaustive search of literature, including Indian sources, often not available to most Indians.

The North-South Divide

While discussing about the commercial uses of biodiversity in Chapter 2, it was stated that the general flow of international trade in the biological resources is broadly directed from the resource-rich and technology-poor 'South' to the biodiversity-poor and technology-rich 'North'. In the context of the international trade of these products, it was established that the available data also corroborate this fact. Another point of concern observed was that the industrialized countries are relatively biodiversity-poor, while developing countries, although technology-poor, are beginning to realize that they are the stewards of the bulk of the earth's biodiversity. This North–South divide is well demarcated till today despite the hard negotiations that have been taking place at global level under the WIPO, UNCTAD, the Secretariat of CBD and at various other forums.

It is at times very disappointing that, if looked at critically, the basic tenets of the CBD and the Articles of the Agreement on TRIPS exhibit that these two treaties do not meet the end goals of the technology-rich 'North' and the biodiversity-rich 'South'. According to Watal (2001), the Agreement on TRIPS is by far the most wide-ranging and far-reaching international treaty on the subject of IP to date and marks the most important milestone in the development of international law in this area. Many authors believe that the Agreement on TRIPS has several provisions favouring developed countries over the developing ones (Dhar and Chaturvedi, 1998; Ghate et al., 1999). To be more specific, Article 27 (*Patentable Subject Matter*) of the Agreement on TRIPS makes it compulsory for a country to have the tool of patents that would protect innovations in all fields including food, health and biotechnology. The arrangement for protection may include a sui generis system or its combination with patent laws, a requirement of efficacy that suits the developed countries. Highlighting the North–South divide, Watal (2001) states that the reason for diverse views might lie, in part, in the fact that the Agreement on TRIPS was the result of bitter North–South negotiations, reflecting strong economic interests on the part of right-owners as well as those benefiting from weaker levels of protection for IPRs. Watal (2001) further recounts that after many years of its finalization and innumerable treatises on the Agreement on TRIPS, there is still an acute need for more information on the negotiating history and interpretation of its provisions. What Watal observed is true even after two decades, and it is still observed that the developing countries are in the need of sustained support in understanding their obligations on the Agreement on TRIPS and in drafting optimal legislative strategies that could meet their vital national interests and developmental goals.

The dividing line between the developed countries and the developing countries broke out during the Uruguay Round when high and uniform standards for protection of IP were proposed and developing countries were not willing to accept such uniform standards (Gandhi and Patel, 2001). This North–South divide is made more explicit by Secrett (2004): Some 500 companies, the great majority incorporated in the 'North', have grown to master almost two-thirds of world trade. Having manipulated trade agreements to secure unprecedented access and de facto control for their companies over natural resources and commodities produced by other nations, the industrialized countries are leading the charge through the WTO to force developing countries through statutory mandate to open up their relatively protected and nationally owned agricultural, financial and service sectors.

The Council for TRIPS is the body, open to all members of the WTO, that is responsible for administering the Agreement on TRIPS, in particular monitoring the operation of the Agreement (Article 68). The main stated objective of the Agreement on TRIPS (as in *the Preamble*) is 'to reduce distortions and impediments to international trade, taking into account the need to promote effective and adequate protection of intellectual property rights, and to ensure that measures and procedures to enforce intellectual property rights do not themselves become barriers to legitimate trade'. Another objective (Article 7) brings up the outlook of social and economic welfare, and states that the protection and enforcement of IPRs should contribute to the promotion of technological innovations and to the dissemination of technology, to the mutual advantage of producers and users of technological knowledge. There is reference to 'a balance of rights and obligations' also (see Box 4.1).

The commentators on the Agreement on TRIPS have consistently pointed out the importance of Article 7 (*Objectives*) and Article 8 (*Principles*) to the developing countries, although no specific obligation ensues from these provisions. It is contended that these provisions set the framework for the Agreements on TRIPS in terms that are consistent with developing country interests.

Above and beyond the Articles on patents for pharmaceutical and agrochemical products included in the Agreement on TRIPS, which invite most of the present-day debate in developing countries on public health and biopiracy, the transfer of technology and the concepts of mutual advantage are other matters of major concern to the developing countries. Prior to the Agreement on TRIPS, the rules governing the protection of IP at the international level were established primarily by the WIPO Conventions. But unlike other international agreements on IP, the Agreement on TRIPS has a powerful enforcement mechanism, and a country that does not adopt IP laws that are Agreement-compliant may have to go through the dispute settlement mechanism as per the provisions of Articles XXII and XXIII of GATT 1994 (Article 64 of the Agreement).

As a consequence to the above outlook, since its enforcement, the Agreement on TRIPS has received a growing level of criticism from developing countries, academics and NGOs. However, because of the rule-making processes in the WTO, and the technical complexities of the laws in question, anything short of a widespread and intense political opposition is unlikely to decrease the power of the Agreement on TRIPS (NALSAR, undated). Ghate et al. (1999) emphatically remark that India is now a signatory to GATT—with the Agreement on TRIPS as one of its components—which has several provisions favouring the developed countries over the developing ones, and TRIPS requires all member countries to provide for a strong 20-year-long patent protection to processes as well as to products based on both domestic and foreign innovations. India did not have the provisions for product patent for pharmaceuticals and agrochemicals prior to the Agreement on TRIPS, and it is noteworthy that the country is concerned about the emerging international trade regime wherein the pharmaceutical and agrochemical products, based on the indigenous biological resources and the associated TK, are likely to enter in a big way. Provisions for patents for these products were introduced in India, effective from 1 January 2005, vide the Patents (Amendment) Act 2005.

In the light shed by the above reflection, the Agreement on TRIPS obviously is the most comprehensive international instrument on IPRs as it aims at setting standards on both the availability of rights and their enforcement in the member countries. Its compliance calls for making new legislation altogether in addition to the revision of laws in respect of civil, criminal and administrative procedures. Also, developing the institutional infrastructure to implement the Agreement on TRIPS in developing countries involves substantial costs (Correa, 1999) as it calls for redefining the role of police and custom authorities under the new regime. Some of the discrepancies are highlighted in the following section.

Limitations of the Agreement on TRIPS

The most alarming part in the Agreement on TRIPS that concerns the developing countries is that it contains no provision preventing a person to claim patent rights in one country over genetic resources that are under the sovereignty of another country. In particular, it contains no provision allowing a country's claim to enforce fair and equitable sharing of benefits from the patenting of its own genetic resources abroad. In the absence of clear provisions providing for a mutually supportive relationship of the Agreement on TRIPS with the obligations under the CBD, implementation of the Agreement on TRIPS may allow for acts of biopiracy. With a view to avoid these conflicts, an amendment in the Agreement on TRIPS to accommodate some essential elements of the CBD is considered necessary.

Under Article 39 of the Agreement on TRIPS dealing with the provision for protection of undisclosed information, members are obliged to ensure protection of undisclosed information through systems developed through appropriate legislations. In the context of the protection of TK, which is major concern of the biodiversity-rich tropical countries, Nair (2001) rightly questions: Considering the vast repository of undisclosed knowledge, practices and products possessed by large number of our *vaidyas*, hakims, artisans and artists, can India bring in a sui generis system of protection of undisclosed information and trade secrets? India along with some other developing countries has suggested suitable changes in Article 39 with a view to provide protection of its genetic wealth and the TK associated with them. The protection of these types of knowledge, and also the resources, would be impossible without deliberating hard on these issues. Nair (2001) argues that even though the Agreement on TRIPS does not define trade secrets, it includes provisions for the protection of trade secrets under the general term 'undisclosed information'. On the other hand, it is silent on the modalities of achieving this and has left it to the individual members as to how to provide such protection. The instruments available in India in this context are the recourse to the breach of contract provisions, and the route of trade secrets, though it would be easier said than done when a large repository of knowledge and practices is vulnerably held by our traditional *vaidyas* and hakims.

The way ahead is not easy for winning advantages from these negotiations, for the ambiguities in the Agreement on TRIPS are probably premeditated. Referring to this as 'constructive ambiguity' that each side of the technology-rich 'North' and the biodiversity-rich 'South' may interpret according to their own convenience, Watal (2001) affirms that the North–South conflict at the time of negotiations on the Agreement on TRIPS was resolved through this 'constructive ambiguity' only. It may be asserted without a pinch of doubt that the most intricate provision in the Agreement on TRIPS influencing the interests of the biodiversity-rich tropical countries is Article 27 (*Patentable Subject Matter*). The intricacies have been scrutinized in the following paragraphs.

Dialogue on Article 27 of the Agreement on TRIPS

The key provision under Article 27 of the Agreement on TRIPS related to the patentable subject matter offers huge protection to the technology-rich 'North', putting the genetic resources and the associated TK of the biodiversity-rich 'South' at the risk of biopiracy: *Patents shall be available and patent rights enjoyable without discrimination as to the place of invention, the field of technology and whether products are imported or locally produced* (see Box 4.2).

On the other hand, the CBD has provisions, though seemingly soft-worded, for equitable sharing of results of R&D and the benefits arising from the commercial use of these resources. A realistic assessment of this situation indicates that the above provision under the Agreement on TRIPS may not remain tenable for long if the truth of the biodiversity-rich 'South' is believed. To test the efficacy of this provision in the light of the three main objectives of the CBD, the benefit sharing arrangements being practised in six very important case studies (the INBio–Merck collaboration in Costa Rica, the ICBG Project in Suriname, the FIRD-TM in Nigeria, the arrangement of benefit sharing with the Knai people in India, the San Hoodia case from South Africa and the argan oil case of Morocco) were examined closely (see Chapter 6). It was found that one marked approach in general was to provide economic incentives in all the cases of bioprospecting, some of which should surely go to the indigenous communities.

It is principally affirmed from the overview of these case studies that the indigenous and local communities must have full usufructs over adequate natural resources, located at an accessible distance, so that their basic needs are fulfilled (Singh et al., 2000). In the developing countries, they largely have continued access to these resources as a right, though undoubtedly the state is the ultimate owner of all

Box 4.2:

The Agreement on TRIPS: Article 27 (Patentable Subject Matter)

1. Subject to the provisions of paragraphs 2 and 3, patents shall be available for any inventions, whether products or processes, in all fields of technology, provided that they are new, involve an inventive step and are capable of industrial application. Subject to paragraph 4 of Article 65, paragraph 8 of Article 70 and paragraph 3 of this Article, patents shall be available and patent rights enjoyable without discrimination as to the place of invention, the field of technology and whether products are imported or locally produced.

    (The terms 'inventive step' and 'capable of industrial application' may be deemed to be synonymous with the terms 'non-obvious' and 'useful', respectively.)

2. Members may exclude from patentability inventions, the prevention within their territory of the commercial exploitation of which is necessary to protect *ordre public* or morality, including to protect human, animal or plant life or health or to avoid serious prejudice to the environment, provided that such exclusion is not made merely because the exploitation is prohibited by their law.

3. Members may also exclude from patentability:

    (a) diagnostic, therapeutic and surgical methods for the treatment of humans or animals;

    (b) plants and animals other than microorganisms, and essentially biological processes for the production of plants or animals other than non-biological and microbiological processes. However, members shall provide for the protection of plant varieties either by patents or by an effective sui generis system or by any combination thereof. The provisions of this subparagraph shall be reviewed four years after the date of entry into force of the WTO Agreement.

Source: www.wto.org/english/docs_e/legal_e/27-trips.pdf (Agreement on TRIPS).

resources and the ultimate right has to apply to the state, as endorsed by the CBD too. As regards the sharing of commercial benefits accruing from these resources with the industries and the corporations, the fundamental point is that it is not only the resource that the indigenous and local communities have the first right to but also the application of the TK, associated with the resource, that is usurped by the corporations, resulting into much higher returns for them as compared to the use of genetic resources without the use of TK.

It has been mentioned earlier that the application of TK leads to a higher hit-rate in screening of the active ingredients as compared to the random sampling method (Gupta, 1999a). The other contention, based on the fact of socio-economic dependence of the indigenous people on these resources, is that unless an alternative for subsistence and income is provided to these people, it would be inappropriate to deprive them of the share in the benefits accruing from the very resource and the associated TK held principally by these people.

The resoluteness of the provision under Article 27 of the Agreement on TRIPS will depend upon the bargaining power and the unity of purpose of the developing countries that they are able to draw from the CBD. The biological diversity that resides on their land masses is, instead of being treated as common property of the indigenous communities, recognized now as the common heritage of the whole of humankind. In the context of ownership issue of the natural resources, one distinct change during recent years has come up—the emphasis now is on the State to have the ownership of the rights over the biological resources, and the authority to determine access to these resources, including the authority to make necessary laws to this effect, is the State only (as provided in the CBD). Though simultaneously there is stress on biodiversity conservation, including the protection of TK, innovations and practices of indigenous and local communities and equitable sharing of benefits accruing from these resources, the people (the indigenous and local communities) still remain at the receiving end and they cannot own the biological resources belonging to their surroundings. In more explicit words, it amounts to saying that the right of ownership of the genetic resources, which ought to rest in the people, rests in the State, and it is now left to the wisdom of the State how far it allows this right to be exercised, in practice, to an extent that the objectives of the equitable benefit sharing, as mandated under the CBD, are met.

It is heartening to gather that in many countries, the rights of the local communities are now being recognized over the biological resources that occur within their habitat, and many institutions, NGOs, authors and thinkers strongly advocate about these rights being intricately connected to the livelihood concerns of these people, though how far these rights can be exercised as a matter of fact has yet to established. The tribal communities in India had customary claims to dependence on forests for their livelihoods, yet an insignificant number of community claims for user rights were being registered in most parts of the country, and though efforts to establish community rights over forest resources began much before the GOI enacted the Scheduled Tribes and Other Traditional Forest Dwellers (Recognition of the Forest Rights) Act or the 'Forest Rights Act' of 2006, the challenge of designing ways to promote individual as well as community rights of forest dwellers to forest resources and assets was taken up as a priority agenda after this Act came into force (UNDP, 2012). Simultaneously, it is also very heartening to learn that the clash between the developed world and the developing countries may get resolved through the intervention of institutions such as WIPO, though it is feared it may take many more years to resolve.

If in the above background the Third World countries adopt a common approach in negotiating with the developed world, an acceptable equilibrium may develop in such relationships early, despite the language in the CBD not being mandatory on these issues. Article 27.2 of the Agreement on TRIPS recognizes that states can exclude from patentability inventions, the prevention of whose commercial exploitation is necessary to protect human, animal or plant life or health or to avoid serious prejudice to the environment. It follows that an invention, the commercial exploitation of which is permitted by law, can never be excluded from patentability. In other words, if commercial exploitation of the natural resources of a country is allowed under any arrangement, then the invention related to that cannot be excluded from patentability. Kothari and Anuradha (1997) conclude that developing non-monopolistic alternatives to patenting, while retaining the right to commercial exploitation, does not seem possible. Continuing on these lines, the issue is discussed further in the following section keeping the mandate of the CBD in view.

To be precise, it may be stated here, before we evaluate and compare the mandates of the Agreement on TRIPS and the CBD, that the arrival of WTO is an announcement of the arrival of the globalization, and the Agreement on TRIPS is the vehicle that begins this process in the new IPR regime. The Agreement on TRIPS, as is evident from its very first sentence in its Preamble, basically aims to reduce distortions and impediments to international trade and to ensure that the procedures to enforce IPRs do not themselves become barriers to trade. This broad framework, together with the provision under Article 41 (*IPR enforcement not to be unnecessarily complicated or costly and not to entail unwarranted delays*), may constrict the domestic laws related to the protection of TK. The objectives of the Agreement on TRIPS (*Article 7*) make things clear by linking the protection of IPRs with the technological innovations and dissemination of technology conducive to socio-economic welfare, yet it may not be of use unless further interpreted in favour of the biodiversity-rich and technology-poor 'South'.

It is also feared that the Agreement may impair the microbiological diversity of the tropical countries because of the protection accorded to microorganisms, non-biological and microbiological processes and even to the plant varieties (*Article 27.3b*). Similarly, Article 29 (*Conditions on Patent Applicants*) of the Agreement requires patent applicants to disclose the specifications sufficiently clear and indicate best mode for carrying out the invention, yet it is not mandatory for the applicant to disclose the source of origin of the biological resource used or admit the use of associated TK that might have provided the lead to the invention (see Box 4.3). A disclosure clause is in great demand on the part of the tropical countries, and is being included in the sui generis legislative systems of most of these countries.

Mandate of the CBD vis-à-vis the Agreement on TRIPS

It is oft stated that the objective of fair and equitable sharing of the benefits arising out of the utilization of genetic resources with indigenous communities may not be achievable unless the IPR laws are necessitated to have stringent norms about the disclosure of the country and the community from which a patentable subject matter and information regarding its application were obtained, and also about the proof of the PIC. Both these requirements are mandated by the CBD, whereas the Agreement on TRIPS is silent on this issue. The

Box 4.3:

The Agreement on TRIPS: Article 29 (Conditions on Patent Applicants)

> 1. Members shall require that an applicant for a patent shall disclose the invention in a manner sufficiently clear and complete for the invention to be carried out by a person skilled in the art and may require the applicant to indicate the best mode for carrying out the invention known to the inventor at the filing date or, where priority is claimed, at the priority date of the application.
> 2. Members may require an applicant for a patent to provide information concerning the applicant's corresponding foreign applications and grants.
>
> **Source:** www.wto.org/english/docs_e/legal_e/27-trips.pdf (Agreement on TRIPS).

negotiations that concluded into the formulation of the Agreement on TRIPS probably did not allude to the objectives or the principles of the CBD, the treaty Dutfield (1999) recognizes as the first one to acknowledge the vital role of TK, innovations and practices in biodiversity conservation and sustainable development as well as the need to guarantee their protection, whether through IPRs or through other means.

This silence may not be of much concern, as the Committee on Trade and Environment under the WTO has stated that the silence of the Agreement on TRIPS in this regard would not preclude bilateral arrangements between countries and companies to ensure such sharing, provided these are compatible with it (Kothari and Anuradha, 1997). Drawing from the striking shift that the CBD makes from unfair exploitation to a legally binding system of exchange, many developing countries wanted to have a full protocol on ABS, being quite aware that it was not sufficient to regulate access as countries of origin. Once natural resources and associated TK leave the country of origin (the resource provider), it is necessary for the country of the bioprospectors and subsequent users (the resource user) to monitor and ensure that the rights of the resource providers (the countries of origin including the respective local communities) are not violated.

There is another aspect that is of concern indeed: the need felt for the countries of origin to cooperate in developing benefit sharing principles, terms and mechanisms that could happen to be a common basis for access regulations. This would be a safeguard against contractual negotiations that exploited two competing resource providers, especially when the countries had common biological resources. This would reassure developing countries in collectively benefiting from a truly fair and equitable benefit sharing system.

The provision under Article 15 (*Access to genetic resources and the equitable sharing of benefits*) of the CBD attempts to define the relationship between providers and users of biological and genetic resources, by laying down the following fundamental principles:

- Recognition of nation states' sovereign rights over biological resources;
- Facilitation of access to genetic resources, for environmentally sound uses, subject to PIC; and
- Fair and equitable sharing of the results of R&D and the benefits arising from commercial or other use of genetic resources.

This stipulation makes the provision contained in Article 8j of the CBD—that the CPs are to protect and encourage the customary use of biological resources in accordance with traditional cultural practices that are compatible with conservation or sustainable use requirements—more comprehensive. In 2002, at the sixth meeting of the COP (COP VI) to the CBD, the Bonn Guidelines on ABS were finalized and adopted. The guidelines were intended to help the CPs, governments and other stakeholders in developing overall ABS strategies, specifically when establishing legislative, administrative or policy measures on ABS and/or when negotiating contractual arrangements for ABS. The guidelines identified the steps in the ABS process, with an emphasis on the obligation for users to seek the PIC of providers. They also identified the basic requirements for MAT and defined the main roles and responsibilities of users and providers and stressed the importance of the involvement of all stakeholders (SCBD, 2002).

The Ad Hoc Open-ended Working Group on ABS (Working Group on ABS) set up under the CBD had worked on these guidelines. However, developing countries in particular were of the view that these guidelines were inadequate, even with the amendments made at COP VI. This issue was taken up at the World Summit on Sustainable Development (WSSD) in Johannesburg in 2002. After long negotiations, it was agreed in Paragraph 42(o) of the WSSD Plan of Implementation to negotiate within the framework of the CBD, bearing in mind the Bonn Guidelines. A number of key developing countries wanted an international legally binding instrument on benefit sharing that would create legally binding rules and procedures that safeguard the rights of provider countries and ensure that user countries fulfil their obligations.

The approach is endorsed in general by several international institutions. For example, the Center for International Environmental Law (CIEL)[2] supports such negotiations that are conducted under the auspices of the CBD, and explicitly address the use of patents that facilitate the misappropriation of genetic resources and associated TK, and that safeguard policy space for varying national systems within developing countries that reflect different national objectives and levels of development (CIEL, 2002).

As regards the Bonn Guidelines, though they were adopted by 180 countries, they are not legally binding. Again, while the CBD encourages source countries to promote access to their biological resources in a regulated manner, the responses to this mandate have been varied. Many countries have been slow in the beginning to develop transparent systems for regulating access and to designate authority to regulate access to a specific government office. Only a few countries, in addition to India, most notably Costa Rica and the Philippines, passed enabling legislation specifically intended to regulate access. Now of course the process has gathered momentum and many more countries have formulated necessary regulations. In the absence of a transparent system for obtaining PIC, usually through a permitting process, negotiating permission to operate and a system for sharing benefits can be complex and difficult (Miller, 2003).

However, agreements have been happening without the regulation, and another outcome has been the progress of biotechnology along with the escalating hunt for genes and associated with it the patent claim over life forms, that have been recognized in the Agreement on TRIPS. The most significant development was the legitimization of a

---

[2] The CIEL with offices in Washington DC and Geneva and staff of international attorneys, working since 1989 to strengthen and use international law and institutions, aims to protect the environment, promote human health and ensure a just and sustainable society. Website: www.ciel.org (Accessed in September 2014).

patent on a genetically engineered microorganism by the US Supreme Court in the famous Chakrabarty case in 1980. The result is Article 27.3b of the Agreement on TRIPS that requires the WTO members to allow patenting of microorganisms in their national laws.

As regards the legal mechanism that could ensure access to and transfer of technology relevant to conservation or making use of biological diversity, or to ensure an equitable sharing of benefits arising out of the use of genetic resources, or to protect the TK of indigenous and local communities, McManis (2004) realistically recounts that the CBD amounts to little more than a declaration of good intentions, and interested parties will need to look, rather, to the IPR regime mandated by the Agreement on TRIPS, together with associated contractual mechanisms, as a starting point.

An account of the assessment as regards the limitations in the protection of TK in the new IPR regime, as mandated by the Agreement on TRIPS, has been attempted in the previous chapter, and the issue of contractual mechanism widely used in many bioprospecting projects may be discussed here now.

In a further discussion related to the implementation of the Agreement on TRIPS by the developing and the least developed countries, Thorpe (2002) asserts that the CBD and the growing recognition of the potential value of biological resources have led to suggestions that the patent system should be more supportive of not only the CBD but also the rights of countries and communities supplying biological resources. Thorpe also affirms the suggestion that patent applicants should be required to disclose the information about the source of any material or associated TK that constitutes a significant part of the invention to be protected. He states that the EU legislation on biotechnology inventions encourages applicants to provide such information, though failure to do so does not prejudice any patent rights.

Review of the Agreement on TRIPS

The North–South divide discussed in an earlier section of this chapter is interpreted more distinctly in the light of two different claims with respect to the Agreement on TRIPS and the CBD: the Agreement on TRIPS is construed to be disfavouring the developing countries, for it is silent about the origin of the genetic resource and associated TK that are commercially used, while the CBD is thought to be favouring them with its objectives of 'the conservation of biological diversity, the sustainable use of its components, and the fair and equitable sharing of the benefits arising out of the utilization of genetic resources' (*Article 1*), and the provision for each CP to 'respect, preserve and maintain knowledge, innovations and practices of indigenous and local communities and promote their wider application with the approval and involvement of the holders of such knowledge, innovations and practices' (*Article 8j*). The attempt on the part of the developing countries has been to pursue at the international forum the review of the Agreement on TRIPS for incorporating suitable amendments with the support of the provisions laid down in the CBD.

Ruiz (2002) believes that ways need to be devised to ensure that the Agreement on TRIPS and the CBD are mutually supportive, even if their objectives point to different overall goals, and with regard to patents in particular, the Agreement on TRIPS should be amended, at the very least, to ensure that when patents are granted over biologically derived inventions that might incorporate TK, the granting procedures require applicants to provide evidence showing that:

- the materials were accessed legally,
- TK when, and if used, was obtained, used or incorporated into the invention with the consent of the corresponding knowledge holders, and
- there exist MAT for benefit sharing.

Sahai (2003) also suggests on similar lines for the amendment in the Agreement on TRIPS to provide for the following in the case of applications for patents relating to biological materials or to TK:

- disclosure of the source and country of origin of the biological resource and of the TK used in the invention,
- evidence of PIC, and
- proof of provision for fair and equitable benefit sharing.

Sahai (2003) further suggests that India should take the position that the rights of farmers and local communities have been reiterated in other conventions, notably the CBD and the International Treaty on Plant Genetic Resources (ITPGR), and these must be reflected in the

WTO. It is for this reason that India has been advocating the linkage between the CBD and the WTO in its submissions to the TRIPS Review Council since the very beginning. Someshwar Singh (1999) very precisely states that a number of developing countries (India, Kenya on behalf of the African Group and some of the Latin American countries) have formulated proposals at the WTO, in terms of the review of Article 27.3(b) and the review of the relevant provisions of the Agreement on TRIPS, rules for the protection of rights of indigenous peoples and their folklore and knowledge. Singh (1999) also recounts that some leading industrialized countries such as Canada have tried to shift consideration of this to the WIPO on the ground of NGOs having better access to it, even as the developed countries try to use the secretive WTO talks to strengthen the rights of their corporations in patenting.

In another citation, the WTO document (WTO, 2001) mentions that two reviews have been taking place in the Council for TRIPS, as required by the Agreement on TRIPS: a review of Article 27.3(b) that deals with patentability or non-patentability of plant and animal inventions, and the protection of plant varieties, and a review of the entire Agreement (required by Article 71.1). The Doha Ministerial Declaration of 2001 says that work in the TRIPS Council on these reviews or any other implementation issue should also look at: the relationship between the Agreement on TRIPS and the CBD and the protection of TK and folklore. Making a note of this development, McManis (2004) expresses his optimism when he mentions that in return for the commitment on the part of developing countries to conserve, sustainably use and ensure access to their genetic resources, and to strengthen legal protection and enforcement of IPR in the developing world, industrialized countries may find it necessary to adopt legal measures to ensure that developing countries and indigenous and local communities located therein will equitably share in the benefits arising out of the use of the genetic resources.

Evaluating the relationship between the Agreement on TRIPS and the CBD, Kothari and Anuradha (1997) assess that in view of the mandate of the COP decisions, the issues that need to be addressed include:

1. The general proposition that the provisions of the CBD should prevail over the GATT agreement, where the principles of conservation, sustainable use and the sharing of benefits arising from the use of biodiversity are in question, needs to be examined carefully.

2. Both national and international actions taken as a follow-up to the GATT provisions, including the Agreement on TRIPS, should be monitored vis-à-vis the impacts of such developments on the objectives of the CBD.
3. To facilitate the realization of objectives of the CBD, such as that of equitable benefit sharing, the existing IPR model under the Agreement on TRIPS should specify that norms of disclosure pertaining to an IPR application should reveal the country of origin and the community that provided the knowledge about the resources pertaining to the patentable subject matter.

These and similar other numerous observations about the vulnerability of the genetic resources that originate in developing countries lead to mixed opinions about the effectiveness of the provisions in the CBD. According to McManis (2004), in contrast to the binding and enforceable provisions of the Agreement on TRIPS, the CBD is essentially a toothless declaration of good intentions, as no enforcement mechanism is specified and much of the treaty language is hortatory rather than mandatory. It is also believed simultaneously that the CBD could take care of the interests of developing countries. Quoting the provision of access to genetic resources being subject to PIC of the CP providing such resources, under Article 15.5 of the CBD, Ghate et al. (1999) opine that the CBD to a great measure reflects the worldwide concern for preventing unfair exploitation of the genetic wealth and TK of the developing countries by the developed world.

Well, would it be enough to have treaties that merely have some reflections on how to make an equitable commercial use of the wealth of biological diversity of tropical countries? Is an equitable sharing of benefits possible in the absence of legal international instruments as opposed to the robust mandate of the WTO? The real optimism would be that as a finality of the negotiations going on at global level, the member countries reach to an acceptable framework conforming to the three main objectives of the CBD. Anything short of that would not serve the desired purpose. It is possible to envision an equitable sharing of the benefits accruing from the commercial use of biodiversity and channel them to promote TK as well as the conservation of biodiversity, taking advantage of the opportunities created by the CBD. In this connection, while the CBD asserts that biological diversity is a global public good, Barrett (1994) goes further to state that countries can be made better off by cooperating because biodiversity exhibits certain global public good characteristics, partly because the existence of biological diversity is valued and partly because the information embodied in biodiversity is valued. Suggesting a way out, Barrett points out that the 'biodiversity problem' is largely one of transferring resources from developed to developing countries in order to compensate the latter for the cost of conservation in their territory. Indeed, it is obvious that the CBD has promising stipulations as regards the equitable distribution of the benefits accruing from the use of the genetic resources and yet it lacks effectiveness, and the rough road of the much talked about globalization demands tough negotiations at the international level in order to synchronize the interests of the developing and the developed countries.

As an obvious consequence, the misgiving about the effectiveness of the provisions in the CBD leaves the developing countries in a dilemma and eventually the benefit sharing policy is left to be defined only in their national laws. Lacking basically in comprehensible legal rights to the local communities over their knowledge or genetic resources, the treaty generally would provide a broad framework to guide ABS arrangements for biological and genetic resources, and more formal instruments such as contracts and MTAs would be needed to agree to a proper basis for regulation of these arrangements.

5

The Access and Benefit Sharing Regulations: Global Scenario

In the backdrop of the implications of the WTO, especially the Agreement on TRIPS globalizing the patent laws, and the tenets of the CBD affirming the conservation of biodiversity, sustainable use of its components and fair and equitable sharing of the benefits accruing from the use of genetic resources, a binding arose on the part of the governments of the biodiversity-rich countries to have their own sui generis system to safeguard their genetic resources as well as the TK held by their indigenous communities. Naturally, it does not mean that the technology-rich and biodiversity-poor countries (developed countries) do not need or should not have any such laws, yet the perspective in those countries is different. The regulations are confined in general to the articles of their patent laws providing for information on the geographical origin of patentable subject material, particularly to protect privately owned resources. As regards the USA, a system of ABS pertains only to the national parks system (Medaglia et al., 2012). The analyses of the biodiversity prospecting partnerships (see the case studies reviewed in Chapter 6) also indicate that having robust ABS regulations for any region or a country may prove greatly advantageous in negotiations for effective bioprospecting partnerships.

India being a signatory to the CBD and also to the WTO was one among the first of many biodiversity-rich countries to have the law enacted for the purpose. While India's Patents Act of 1970 was suitably amended to effect changes in its provisions in conformity with the provisions of the Agreement on TRIPS as well as the basic objectives of the CBD, the Biological Diversity Act of 2002 came into effect on 5 February 2003. This Act has its own sui generis system for protection of the genetic resources of India, and it also comprises all the necessary regulations related to ABS issues. The Patents Act of India complements this provision of the Biological Diversity Act of 2002 by making it mandatory for the applicant of a patent to submit a declaration under Form-1 (*Application for Grant of Patent*) of the Patent Rules 2003 to the effect that 'the invention as disclosed in the specification uses the biological material from India and the necessary permission from the Competent Authority shall be submitted by me/us before the grant of patent to me/us' (GOI, 2012).

The objective of India's Biological Diversity Act is in consonance with the three basic tenets of the CBD, that is, to provide for conservation of biological diversity, sustainable use of its components and fair and equitable sharing of the benefits arising out of the use of biological resources and TK associated with these resources. The unique feature of the Biological Diversity Act of 2002 is its provision for establishment of NBA at national level to accord approval, on application, to obtain a biological resource from the country, including the knowledge associated with the resource for research or for commercial utilization. The approval is obligatory for foreigners, non-resident Indians and corporate bodies or any other organization not registered in India. Prior approval of the NBA is also required for transfer of the results of research relating to biological resources occurring in India or obtained from India for commercial purpose to any foreigner or non-resident Indian. For the commercial use of biological resources by Indian citizens within the country, giving prior intimation to the State Biodiversity Board (SBB) concerned is adequate.

Apparently, the enactment of India's Biological Diversity Act of 2002 is more of an outcome of the post-1995 IPR regime and less of the post-CBD regime, as the Agreement on TRIPS for sure opened the floodgate of misappropriation of the knowledge and resources of indigenous and local communities. The Act does not forbid patents granted globally on genetic material that originates from India; it only regulates the bioprospecting operations in a way that there is an equitable sharing of benefits accruing from the commercial use of India's biodiversity.[1] In the context of the North–South divide and the relationship between the Agreement on TRIPS and the CBD, India as the signatory to both of the treaties has emerged, partnering with Brazil and South Africa, as a campaigner of the interests of the developing countries. The CBD came into force in December 1993 and the WTO came into existence with effect from January 1995, but the campaign for equitable sharing of benefits accruing from the tropical biodiversity, which began in late 1990s, is still going on. The developing countries have been dwelling upon suitable changes in the Agreement on TRIPS at different forums for more than a decade now, but no appropriate resolution has been arrived at. Not the least of anything has come out so far that one could term as final from the constant negotiations and deliberations. Of course the countries, for example, India, South Africa, Brazil, Costa Rica, Peru, Philippines and Colombia, have already enacted suitable laws for ABS regulations, paving the way for protection of TK and an equitable sharing of benefits accruing from the commercial use of genetic resources of tropical countries.

The ABS Regulations and Biodiversity Laws

Let us have a look at the global scenario of ABS legislations in order to have a better concept of the complexities involved. We can pertinently begin with mentioning about some biodiversity-rich countries that could not develop their biodiversity laws due to some limitations but still are managing the affairs. Mexico is one—the southern North American country having mega biodiversity is hindered by the conflicts of land tenure and resource use in rural areas, and could not enter the ABS regime (Medaglia et al., 2012). But the Mexican Criminal Code having provisions for imprisonment and fine for illegal collection and traffic of genetic resources is effective. Some other laws also

---

[1] India is one of the 17 mega biodiversity countries of the world with only 2.4 per cent of the land area, and yet, with over 46,000 species of plants and 81,000 species of animals recorded in the country so far, accounting for 7–8 per cent of the recorded species of the world. *Source*: www.nbaindia.org/ content/ 19/16/1/ faq.html (Accessed in November 2014).

are helping the country's natural wealth, though IPR laws still need legislative initiatives. Morocco, the north-western African country, also does not have any ABS regulations. The country is rich in forest and marine ecosystems and is home to the 'liquid gold', that is, argan oil, providing a lead to bioprospecting and bringing in the sharing of benefits through 'CSR' approach (Robinson and Defrenne, 2009).

In the context of ABS regime, Costa Rica is a great example, as it has one of the most advanced and transparent regimes with clear and well-defined regulations for obtaining access to genetic resources, and as a result, there are a significant number of contracts with companies in the life sciences industry (BIO, 2013). The Biodiversity Law of Costa Rica includes, in addition to the provisions for decentralized funds, exchange of components of biodiversity as well as aim to achieve the understanding of the value of biodiversity. It also has a clear provision for CIR and even involvement of the private sector and civil society in order to promote investment for the commercial use of biodiversity. Some of the significant features of Costa Rica's Biodiversity Law, formulated as long back as 1998, are listed below:[2]

- The Article on international cooperation (Article 12) tells that the State should promote, plan and orient national activities, foreign relations and cooperation with national neighbours with respect to the conservation, use and exchange of components of biodiversity present in the national territory and in the trans-frontier ecosystems of common interest. Likewise, it should regulate the entry to and exit from the country of biotic resources.
- The Article on financial instruments (Article 36) provides for decentralized funds that enter the system by any concept by means of trust funds or other instruments, whether they are for the entire system or specifically for each conservation area.
- Costa Rica expressly recognizes and protects, under the common denomination of sui generis CIR (Article 82), the knowledge, practices and innovations of indigenous and local communities related to the use of components of biodiversity and associated knowledge.

[2] Costa Rica's Biodiversity Law of 1998. *Source*: eelink.net/~asilwildlife/costa.pdf (Accessed in July 2014).

- The country envisages to integrate the educational plans of all anticipated levels (Article 86) to achieve understanding of the value of biodiversity.
- The Biodiversity Law also puts across the involvement of private sector and civil society (Article 98) to promote investment for the conservation and sustainable use of biodiversity.

But among all the countries, Brazil is the most distinctive example, because regardless of the ample provisions that the Brazilian federal government has now, which constitute the most stringent instruments for regulating ABS issues, there is no single primary law in force. Brazil, a country that is among the world's few countries that have mega biodiversity, unlike the framework of the comprehensive laws of its partner countries, puts up the most surprising scenario. There have been several initiatives to regulate access to Brazilian genetic resources since 1994, but no law has yet been approved (Medaglia et al., 2012). The Brazilian federal government adopted the 'Provisional Measure' to address the ABS issues, which was followed by creation of Genetic Heritage Management Council (Conselho de Gestao do Patrimonio Genetico or CGEN), a government entity led by a representative of the Ministry of Environment. The CGEN has the mandate of making provisions and implementing the regulatory arrangements for access to genetic resources and TK, rights and obligations, IPR measures and dispute settlement, PIC, MAT for material transfer and benefit sharing including legal regime for benefit sharing contracts. However, the overall legal framework comprises a number of decrees, policies, resolutions and laws besides the Articles of Provisional Measure. The task of the CGEN, being a collective inter-ministerial body stressed with the overachieving objectives and complemented with framework for making non-compliance a punishable act, became too intricate particularly with the monitoring of various aspects of ABS regulations.

Regarding the access authorization from the CGEN, a closer analysis of the legal framework by Biotechnology Industry Organization (BIO)[3]

[3] BIO is a Washington DC-based trade association representing biotechnology companies, academic institutions, state biotechnology centres and related organizations across the USA and in more than 30 other countries. Its members are involved in R&D of innovative health care, agricultural, industrial and environmental biotechnology products. Website: www.bio.org (Accessed in May 2015).

finds that the entire legal process, involving the process of obtaining access to the genetic material up to filing patent applications on technologies and ultimately to commercializing products derived from genetic material obtained, is so complex that it is practically impossible to navigate (BIO, 2013). In its proposal for reform of Brazil's ABS regulations, BIO (2013) mentions that the rules do not specify the information that must be contained in the proposed research and anthropological reports, and there is little guidance as to which institutions are approved by CGEN, and regarding the benefit sharing agreement, it is not perceptible who will be the parties to the agreement. The revision of Brazilian ABS legislation is still underway, and several draft bills have been discussed by stakeholders (Medaglia et al., 2012).[4]

In that respect, Costa Rica and India offer far better examples of the most comprehensive ABS regulations. Costa Rica's Biodiversity Law of 1998 and India's Biological Diversity Act of 2002 singularly comprise all the regulatory provisions for commercial as well as noncommercial use and conservation of biological resources, protection of the associated TK, robust benefit sharing rules, appropriate regulation of bioprospecting and patent regime, basic prerequisites for access and PIC, transfer of technology, participatory approach at the regional and local levels, maintaining inventory of the biological resources and associated TK, fulfilling the objectives of the CBD, validating the rights of indigenous and local communities and their participation in the decision-making process and so on.

Costa Rica also has a very strong institution building record. The country's biodiversity is administered by the MINAE through the National System of Conservation Areas and the National Commission for Biodiversity Management. The Biodiversity Law of 1998 and the formulation of Costa Rica's National Strategy for the Conservation and Sustainable Use of Biodiversity (adopted in 1999), using a highly participatory process at the local and national levels, have facilitated the establishment of the National Commission for Biodiversity Management, which together with National System of Conservation Areas is responsible for the administration of the country's natural resources.[5]

---

[4] Brazil recently has enacted a new biodiversity law (Law No. 13,123 of 20 May 2015, coming into force in November 2015) to regulate the access to genetic resources and associated TK. The new law revokes the earlier 'Provisional Measure' (No. 2186, in effect since 2001).
[5] *Source*: http://www2.inbio.ac.cr/en/biod/bio_quebiod.htm (Accessed in July 2014).

The Philippines also is the country that formulated its Biodiversity Law during the initial years. The Wildlife[6] Resources Conservation and Protection Act, 2001, of the Philippines (*the Republic Act No. 9417 of the Philippines, 2001*) has pretty unambiguous and evident regulations for the commercial use of biodiversity (WIPO, 2001). Section 14 of the Act states that bioprospecting shall be allowed upon execution of an undertaking by any proponent, stipulating therein its compliance with and commitment to reasonable terms and conditions that are necessary to protect biological diversity. The Act has the prerequisite that before granting the permit for bioprospecting, the PIC, in accordance with the existing laws, shall be obtained by the applicant from the indigenous and local communities or their management board or private individual or entity. Section 14 also specifies that if the applicant is a foreign entity or individual, a local institution shall have to be involved in the research, collection and technological development of the products derived from the genetic resources. The Act has penal clauses (Section 28)[7] and also has provision (Section 21) for levying fee, to be determined upon consultation with the concerned groups, for issuing permits for bioprospecting with an additional export permit fee if the deal involves export of the species.

The Philippines also has adopted the fund mechanism for channelling of the funds (Section 29). Named as the 'Wildlife Management Fund' that derives from fees, charges, donations, endowments, administrative fees or grants in the form of contributions, as well as the fine imposed or the damage awarded is aimed at supporting scientific research, enforcement and monitoring activities, enhancement of capabilities of relevant agencies, as well as financing the restoration of habitats.

Among the other countries, Peru and South Africa enacted their Biodiversity Laws in the year 2004. The Congress of the Republic of Peru passed its Biodiversity Law[8] as 'Act on the Protection of Access to Peruvian

---

[6] 'Wildlife' is defined in the Biodiversity Act of the Philippines as the wild forms and varieties of flora and fauna, in all developmental stages, including those which are in captivity or are being bred or propagated.

[7] The sentence of imprisonment up to a maximum of 12 years and/or fine of maximum five million pesos, depending on the kind of violation of the Act and the species (critical, endangered, vulnerable or threatened) involved in such a violation.

[8] *Source*: www.wipo.int/wipolex/en/text.jsp (Accessed in July 2014).

Biological Diversity and the Collective Knowledge of Indigenous Peoples'. The Act establishes a 'National Commission' for the protection of biodiversity and the collective knowledge of indigenous people. The 'National Commission' has a representative of the National Institute for the Defense of Competition and Intellectual Property Protection as its chairperson, and its members are drawn from the Government, other national institutes and councils and also from civil society. The National Commission has to draw its resources from donations and also by way of international cooperation. Its functions have been very well thought of and include the following:

- Establishing and maintaining a register of biological resources and collective knowledge of indigenous peoples of Peru;
- Promoting links with State and civil society regional participatory bodies;
- Establishing permanent information and dialogue channels with the industrial property offices of other countries;
- Identifying and following up patent applications filed or patents granted abroad that relate to biological resources or the collective knowledge of indigenous peoples of Peru, and instituting actions for annulment concerning patent applications filed and patents granted abroad that relate to biological or genetic resources, or the collective knowledge, of indigenous and native peoples of Peru; and
- Drawing up proposals for the defence of the position of the State and of indigenous people of Peru in different international fora with a view to prevent and avoid biopiracy.

The Biodiversity Law of South Africa, though a contemporary law evolving simultaneously with the Biodiversity Laws of Peru, India and few other countries, has a different approach altogether in the treatment of the subject of commercial use of biodiversity. It is unique in the sense that its objectives, which are inclusive of the three fundamental objectives of the CBD (the conservation of biological diversity, the sustainable use of its components and the fair and equitable sharing of the benefits arising out of the utilization of genetic resources), also aim at cooperative governance in biodiversity management and conservation and at achieving these objectives through the support of South African National Biodiversity Institute, a juristic institution established under the Biodiversity Act (Section 10). The Act was enacted by the Parliament of the Republic of South Africa in the year 2004 within the framework of the National Environmental Management Act 1998 of South Africa and was titled aptly as the National Environmental Management: Biodiversity Act 2004.[9]

The Biodiversity Act of South Africa is rendered more comprehensive as its Section 1 includes numerous definitions, including those of the components of biodiversity, the genetic material and the genetic resource in addition to other common terms fairly used the world over. The contents comprising the various clauses of the Act, formulated still more comprehensively through the inclusion of provisions for establishment of various institutions, make the Act unique. For example, the Act (Section 10) provides for the establishment of the 'South African National Biodiversity Institute' with the mandate for monitoring and reporting on a regular basis the status of the country's biodiversity, including the conservation status of all listed, threatened or protected species, and also the impacts of any genetically modified organisms. The Institute is bestowed with enormous powers and is entrusted to manage all national botanical gardens, establish, maintain and preserve collections of plants in national botanical gardens and in herbaria and also has to undertake research on indigenous biodiversity and the sustainable use of indigenous biological resources. The Institute is to act as an advisory body to the Government on any matter regulated in terms of the Act and on all matters relating to the implementation of the Act or any international agreements related to biodiversity.

For the purpose of biodiversity planning, the Act (Sections 38 and 39) has provision for a 'National Biodiversity Framework' to be prepared for an integrated approach to biodiversity management by organs of state in all spheres of government, NGOs, the private sector, local communities and other stakeholders. As regards the ABS arrangements, if an application is submitted for bioprospecting involving export of indigenous biological resources or any other kind of research, a permit system is laid down in the Act. To seek such a permit, it is mandatory on the part of the applicant to disclose all information concerning the proposed bioprospecting and the indigenous biological resources to be used. In order to safeguard the interests

---

[9] *Source*: www.wipo.int/wipolex/en/text.jsp (Accessed in July 2014).

of the on-site stakeholders (members of the indigenous communities whose knowledge of or discoveries about the indigenous biological resources are to be used), the applicant and the stakeholders must have entered into an MTA and a benefit sharing agreement with the provision for sharing in any future benefits that may be derived from such bioresources.

As in most of the other ABS regulations, for the channelling of funds, South Africa also has a distinct provision for establishment of 'Bioprospecting Trust Fund' in its Biodiversity Act (Section 85), which would have inflow of all the sum arising from benefit sharing agreements and MTAs, and from which all payments to, or for the benefit of, all the stakeholders have to be made. The Act also has penal clauses (Sections 101 and 102), providing fine and/or imprisonment up to the maximum period of five years depending on the violation of specific Section the Act.

Though South Africa's Biodiversity Act is quite comprehensive and robust, another regulation was adopted subsequently, known as the Bioprospecting, Access and Benefit Sharing Regulations 2008.[10] The tenets of the Regulations are based on the fundamental idea regarding bioprospecting, stating (Section 1) that the commercialization phase of a bioprospecting project begins when the research, development or application of a country's biological resources matures into or takes shape of action for potential commercial use or when the industrial exploitation is sufficiently established. The commercialization phase follows the discovery phase of a bioprospecting project. South Africa came up with the detailed ABS regulations only to further regulate the permit system set out in Chapter 7 of the Biodiversity Act in so far as that system applies to bioprospecting involving any bioresource or export of any bioresource for any kind of bioprospecting or any other kind of research. The Regulations describe the application procedures for bioprospecting and explain that an integrated export and bioprospecting permit also may be issued if it is contented that the export of indigenous bioresources will be for a purpose that is in the public interest including the conservation of biodiversity in South Africa, economic development of South Africa and enhancing the scientific knowledge and technical capacity of South African people.

---

[10] *Source*: www.wipo.int/wipolex/en/text.jsp (Accessed in July 2014).

The Regulations also explain the MTAs, the process of ABS arrangements and the administration of the Bioprospecting Trust Fund.

Against the backdrop of the ABS regulations at global level and before examining the finer points of India's Biological Diversity Act as an illustration, we shall review in the following section how the IPR regime is behaving now. This aspect is discussed with special reference to India, where the product patent regime came into force in the year 2005 as per the prerequisite of the Agreement on TRIPS.

Patent Filing in India: Some Indications

General awareness about the significance of IPRs has been increasing in India, and the Indian industry responded positively to the new demands of IP laws (TIFAC, 2000). As highlighted in the *Intellectual Property Rights Bulletin* of the GOI, during the period from 1995 to 1999, an average growth of 23.4 per cent per year was observed in filing of the number of applications for patent in the country, which indicates that the Indian industry, especially the private sector (as they dominate in these figures), responded well to the demands of the IP laws, taking appropriate steps for protection of its research results (TIFAC, 2000). It is noteworthy that pharmaceuticals have always been in the top list of patent filings the world over, and very obviously, while discussing the aspects of commercial use of biodiversity, the trend in patent filings in this group of products is plausible to be examined.

Interestingly, while studying the trend in patent filings in this field, we find that a significant number of pharmaceutical product patents granted in India since 2005 appeared to relate to traditional herbal preparations (Park, 2010). This trend is an important indication that the provisions under Sections 6 and 7 of India's Biological Diversity Act of 2002 are much significant in the post-1995 era, and an effective compliance of these provisions should be able to ensure that the applications for grant of patents wherein the research results relate to the biological resources of Indian origin and are based on TK associated with these resources are not entertained unless an equitable sharing of benefits becomes the central part of the process. There is another indication that the new trend involves patent applications from foreign companies to a larger extent. For example, in the year 2010, out of 185 pharmaceutical product patents granted in India, about 38 per cent were granted to the USA while India could bag only about 10 per cent (Rau et al., 2012).

The IPR regime in the pharmaceutical sector is deepening its roots more and more, and some new studies validate the outlook mentioned in the above paragraph. As Rau et al. (2012) have found, under the category of drugs, Indian patent office granted patents ranging below 500 in numbers year by year during the period 1997–98 to 2005–06, while during the period 2006–07 to 2009–10, the year-by-year numbers of patents granted ranged from 530 to 1,207. The study by Rau et al. (2012) concludes that 'the number of pharmaceutical industries filing for patents has to increase drastically in coming years from India'.

If filing of product patents in the pharmaceutical sector boosts, as indicated above, the broad base for the drug discoveries will certainly be formed on the wealth of genetic resources and the associated TK. Now whether the pharmaceutical companies are small or conglomerates and transnational giants, the repository of the biological resources is the same, lying largely in the tropical regions of the world. The question of protection of the TK against 'biopiracy' and through the patent regime brought into force by the WTO is a distant possibility. At this stage, letting the interventions on the part of like-minded countries (primarily India, Brazil and South Africa) continue for suitable changes in the Agreement on TRIPS at international forums, the first and foremost concern should be how to circumvent the provisions of the Agreement on TRIPS at local level in order to protect the TK, and shun the inconsistencies in the equitable sharing of benefits in the domain of biological resources.

Another question is raised about how to provide protection of patents to the formulations and products that were developed over a period of hundreds of years. In this context, Sharma (2000) observes that hundreds of patents are being taken out on TMs the world over, the difference being that while India continues to debate and discuss the issue, other countries are using scientific explanations for the TK and applying for patents. Besides that, what was observed prior to 1995 and much before the product patent regime began in India in 2005, and that is very astonishing too, is that even very successful Indian companies were not patenting their R&D results, and major players in the patenting scene in India have been transnational companies, a trend that is obvious if we scan the patents granted in India between 1972 and 1994 (Ganguli, 1998). Now the question is: Do we foresee that the new product patent regime and the strict ABS regulations are going to bear the fruits India anticipates? The deliberation in the following sections includes this aspect.

India's National Biodiversity Authority

The Biological Diversity Act of 2002 is a significant Indian legislation in connection with the two important provisions of the CBD: one that recognizes the sovereign rights of countries of origin over their biodiversity resources, and the other that recognizes the need to share benefits accruing out of commercial utilization of the genetic resources. The Act aims to promote the conservation, the sustainable use and the equitable sharing of benefits of the biological resources including their habitats and microorganisms. Establishment of the NBA at national level, the SBBs in the states and the BMCs at local level offers the institutional means to act out the implementation part of the Act. The NBA with its headquarters in Chennai is the foremost central agency under the GOI for implementation of the Act, while the SBBs are the state agencies with similar mandate. As regards the BMCs at local level, in order to entail optimum advantage from the use of natural resources of smaller regions and engage as well as empower local knowledge groups and government officials, the Act is quite flexible about the constitution of the BMCs. It is not necessary that each and every village or taluka or block or district has a BMC. The NBA (2013a) specifies that 'potential areas rich in biodiversity and locations where there is popular interest or support should be identified and BMCs established'.

The flexibility within the overall umbrella of the NBA ought to be stupendous as India is a big country and several states in India have their own resource management institutions and enactments, involving local hill management councils, joint forest management committees, Van Panchayats and so on. The local traditions as well as traditional rights and concessions also are varied. Even the conflicts innately arising from the enactment of the Scheduled Tribes and Other Traditional Forest Dwellers (Recognition of Forest Rights) Act (or the Forest Rights Act) of 2006 bring to the fore the issue of right of ownership over the products that are gathered from the forest areas by traditional forest dwellers. Obviously, the BMCs would have more varied characteristics, would need more freedom, and it may not be possible to act on a day-to-day basis in a centralized structure. Obviously, the road to the implementation of the ABS regulations is not so smooth and the India Patent Offices located in Delhi, Mumbai, Kolkata and Chennai, the NBA headquarters located in Chennai or the state headquarters of the SBBs may seem to be too far from the BMCs. The future expectations would lie more in the SBBs and the effectiveness of the relationships between them and the BMCs within their jurisdiction. The everyday communication also needs to be in the local vernaculars. On the part of SBBs, for mass coverage and for more penetrating approach, awareness campaigns should be in place and the guidelines and regulations, approach papers and resource data including PBRs also should be in vernacular. Dr K. Kasturirangan, member of India's top planning body, the Planning Commission, very aptly emphasized in his keynote address to the officials of the SBBs in one of its meetings that the SBBs are key to the implementation of the Act and the Act's effectiveness of implementation depended on the capacities, awareness and interests of SBBs. He stressed that strengthening of the institutional mechanism of the SBBs was necessary (NBA 2013b). This aspect is examined in the following paragraphs.

The guidelines for operationalization of BMCs that were issued by the NBA (2013a) proffer that based on the recommendations of the SBB concerned, the state government shall designate Nodal Officers for each district to oversee the various matters of the BMCs and shall also provide adequate fund and support staff at nodal levels. These guidelines also hold out that the SBB shall formulate district-level Technical Support Group comprising officials, institutions and people of excellence and expertise in the biodiversity issues at local level, which should guide the BMC in its operations. The whole structure appears just right, as at the implementation level, the SBBs and the BMCs are assigned to have a central role in providing access to the genetic resources as well as in providing the arrangements for sharing of benefits accruing from these resources.

If an assessment of India's Biological Diversity Act of 2002 is made with a view to test its efficacy as regards the ABS arrangements, an apprehension arises that the NBA may, in the course of its regulatory role, end up slowing the process of applying for patents based on research or information developed from Indian biological sources, as the situation is already alarming with the Indian Patent Office having over 30,000 unprocessed applications at a time (EPW, 2002). The road is rough and long and tedious, and any arrangement that makes the process of ABS more complicated shall directly complicate the process of granting IPRs that include the use of the genetic resource or the TK associated with it. This would be contrary to the provisions of the Agreement on TRIPS, as the Article 41.1 provides that members shall have to enforce action against infringements, along with actions deterrent to further infringements, 'applied in such manner as to avoid creation of barriers to trade'.

However, within the national boundaries, the process may not be as complicated as alarmed above. Gangopadhyay and Mohan (2005) highlight very well that the act does not hinder the utilization of biological resource or conducting biosurveys and utilization for commercial utilization by citizens of India who only need to give prior intimation to the SBB. The local people and communities including the cultivators and *vaidyas* and hakims can continue utilizing these resources as before and do not need to intimate the SBB. There should be an enforcement system with well-organized instruments and suitable agencies to check any misuse of the freedom to the local people and communities in utilizing these resources. Professor Madhav Gadgil (2003) observes that the attempt of India's Biological Diversity Act of 2002 is significant in the respect that no international agreement has as yet been arrived at as to how to put into effect the relevant CBD provisions, and the act should help direct proper flows of benefits of commercial uses of biodiversity to holders of TK, as well as to grass-roots innovators.

As regards the commercial uses of the products of biodiversity across the national boundaries, India's Biological Diversity Act (Section 19) provides that approval from the NBA shall have to be obtained by any non-Indian applicant (including non-resident Indians and corporate bodies having non-Indian share in capital or management) who intends to apply for patents, whether in India or outside, or intends to obtain any biological resource or associated knowledge for commercial use. The results of research also, if related to India's biological resources, cannot be exported without approval from the NBA by such applicants. The provision goes beyond the step of granting of permission only, and the IPRs granted outside the country in an illegal manner on any of India's biological resource or the associated knowledge may be opposed through necessary measures taken by the NBA on behalf of the GOI.[11]

The novelty in India's Biological Diversity Act is its penal clause (Section 55) that provides for terms of imprisonment up to five years and/or fine up to ₹10 lakh if anybody undertakes biodiversity-related activities in contravention of the provision of the Act, including transfer of results of research, or applies for IPRs without approval of the NBA. The penal clause also stipulates for terms of imprisonment up to three years and/or fine up to ₹5 lakh if commercial use of a biological resource is undertaken without prior intimation to the SBB, or if any activity that is detrimental to the conservation and sustainable use of biodiversity or contrary to the objectives of an equitable sharing of benefits is carried out.

As would be expected about any new legislation that comes into force, there are also apprehensions about India's Biological Diversity Act. A section of people cite certain structural limitations under the Act, and an apprehension is raised that though Section 3 provides for prior approval of the NBA applicable to any non-Indian individual or a corporate body having non-Indian participation, this restriction shall not apply to collaborative research projects approved by the GOI (Section 5). The bill that brought the draft Act in the Parliament of India attracted sharp criticism on a variety of aspects, especially on the issue of collaborative research projects (EPW, 2002). The particular clause in the bill allows the movement of biological resources otherwise regulated, which may encourage biopiracy and alienate indigenous farmers from their resources. There is also some concern that the NBA may in the course of its regulatory role end up slowing the process of applying for patents based on research or information developed from Indian biological sources. Excessive regulation may also stifle domestic commercial deployment of these knowledge sources, especially since it requires prior intimation to the SBBs, although Indian citizens and companies are allowed free access to use the resources for research purposes within the country (EPW, 2002).

---

[11] Rule 12 (xix) of India's Biological Diversity Rules 2004.

Table 5.1:
Status of the applications received by the National Biodiversity Authority of India

	Total Applications Received (from 2003–04 to 2013–14)	*Applications Cleared (from 2006–07 to 2013–14)*	*Agreement Signed by the Applicant with NBA (from 2006–07 to 2013–14)*
Access of Bioresources for Research/Commercial Purpose	152	37	19
Transfer of Research Results	36	15	12
Applying for IPR	600	391	63
Third-Party Transfer	76	38	23
Total	**864**	**481**	**117**

Source: www.nbaindia.org (Accessed in July 2014).

It is now more than a decade that India's Biological Diversity Act of 2002 came into effect and the NBA came into being, and the apprehensions as noted above may well be put to scrutiny in the light of the efficacy of the recent benefit sharing arrangements approved by the NBA. During the last eight years, the NBA has entered into agreement for 117 cases under the various provisions of the Biological Diversity Act of 2002. The status of applications submitted to the NBA and the agreements signed between the applicants and the NBA is given in Table 5.1. While the agreements signed should provide a good field for study and scrutiny, it is too early to comment on the status of the programmes taken up under these agreements due to lack of further information.

As one would expect, there are limitations of any new institution or a new mechanism that takes shape, and, it may well be stated at this stage that, in principle, the establishment of the NBA provides a framework for the integration of local communities in the process of conservation and resource management. The institutions associated with the issue of biodiversity are certainly found lacking in mutual coordination and cooperation at the local level presently, and it is the observation of this author as an individual, having worked at district level in central India for many years, that acting in isolation is very common in most of the districts of India. Now when one looks at the framework provided under the umbrella of the NBA as per the provisions of the Biological Diversity Act, one would well realize that adopting such a strategy under the NBA that integrates all actors and defines their respective roles through the Nodal Officers designated at district level by the state governments as well as the district-level Technical Support Group would surely be the best practice. The resource use cannot be controlled in the absence of adequate ABS regulations, and this calls for integrated efforts on the part of interdepartmental setups and also the local communities. The routine management practice of the local biological resources need to be seen as a single resource with the local inhabitants at the centre and then set for inter-sectoral coordination and cooperation to promote an equitable sharing of the use of the biodiversity, which should be possible through the BMCs. The NBA fits very well in its role to provide that framework in India, and if the things fall into place as provided in the law, it would be possible to achieve the three basic objectives that are common in India's Biological Diversity Act and the CBD.

The practice of creation of the PBRs might play the central role in this framework, and as regards the protection of TK, the provision under India's Biological Diversity Act (Section 36.5) for 'such measures' that may include registration of TK and the sui generis system (with the liability of documentation of biological diversity and chronicling of knowledge being laid under Section 41.1 on BMCs) has to be put to the test of time, for there are not enough experiences to corroborate that the documentation could be the ultimate tool for the protection of TK. The preparation of PBRs is inscribed as the main function of the BMCs as per Rule 22.6 of the Biological Diversity Rules 2004. As reported, more than 1,100 PBRs have been documented (NBA, 2013b), yet this tool also has to stand the test of time.

Benefit Sharing Mechanism in India's Biological Diversity Act

The provision under Section 6.2 that the NBA may, while according its permission for any IPR for invention based on research or information on India's biological resource, impose benefit sharing fee or royalty that would be deposited in the National Biodiversity Fund (NBF) constituted for the purpose carves out a mechanism for benefit sharing. Section 21 of the Act empowers the NBA to frame guidelines that would outline the framework for determination of equitable benefit sharing. The benefit sharing shall be effected in a combination of the following manners:

1. Grant of joint ownership of IPRs to the NBA, or where benefit claimers are identified, to such benefit claimers;
2. Transfer of technology;
3. Location of production, R&D units in such areas that will facilitate better living standards to the benefit claimers;
4. Association of Indian scientists, benefit claimers and the local people with R&D in biological resources and biosurvey and bio-utilization;
5. Setting up of venture capital fund for aiding the cause of benefit claimers; and
6. Payment of monetary compensation and non-monetary benefits to the benefit claimers as the NBA may deem fit.

The Act also burdens the NBA to draw guidelines through which an individual or group of individuals or organizations holding access to the biological resource or the associated TK may be paid directly as per the agreement arrived at for the purpose. If compared with the principles of the Trust Fund or the Forest People's Fund (FPF), the benefit sharing mechanism under Biological Diversity Act appears to be too untailored. It lays down important aspects such as sharing of benefits in accordance with terms and conditions agreed between the person applying for such approval, local bodies concerned and the benefit claimers, but the very process of benefit sharing is a complex set of negotiations between various stakeholders. At the time of finalizing the actual contractual partnership programmes that include more than one lateral agreements between several agencies such as the funding agency, the research institution, the local community, the state agency, local traditional healers and other local groups, the negotiation process will be tough as to how to channel the fund through the Local Biodiversity Fund, what could be the activities that were to be embodied under such agreements, how the interests of the indigenous and local communities would be taken care of and so on.

If the above mechanism of fund flow provided in the Act is compared with the trust fund-based mechanism of benefit sharing in the case studies described in Chapter 6, it is observed that the trust fund mechanism of benefit sharing is very much a case-specific approach while that under the Act is too broad in nature. At times, it may lead to conflicts as to how to divide the inflow of the fund and redistribute the benefits accruing from a local genetic resource base if the partnership programme involves bioprospecting and is global in nature. The trust funds engage specific agreements, a constitution, a board of management (and/or board of trustees/advisory board), and are mostly registered as per local laws. Therefore, the trust funds would apparently contain flexibility and would suit local customs and traditions. As regards the fund mechanism under India's Biological Diversity Act, the objectives of equitable sharing of benefits may not be feasible unless it effectively conforms to the trust fund model.

Another apprehension is that the centralized system of fund management with a government-based structure, with an obvious 'top to bottom' approach, may lead to skewed sharing of benefits at the local level. However, the provisions such as the one in India's Biological Diversity Act (Section 41.3) that empowers the BMCs to levy charges by way of collecting fees from any person for accessing or collecting any biological resource for commercial purposes may be more effective. Another provision (Section 21.3) lays down that the NBA may also direct that the amount shall be paid directly to such individual, group of individuals or organizations in accordance with the terms of any agreement wherein the biological resource or knowledge was a result of access from specific individual or group of individuals or organizations.

Taking it up further, the approach in such a benefit sharing arrangement should be 'bottom-up', involving the various institutions entering into the contractual partnership and the benefit claimers (indigenous and local communities) engaged in a bioprospecting partnership in the decision-making process. The trust fund mechanism as adopted in various bioprospecting partnerships has been described at length in Chapter 6, and further deliberations on the subject have been laid down in Chapter 7 wherein it shall be observed that the methodology adopted in the bioprospecting partnerships mostly maintain multilateral agreements.

In a nutshell, it may be suggested that an efficient contracting process should be preferred that should base more on the involvement of indigenous communities and local institutions with an inbuilt flexibility rather than on a hard-line bureaucratic setup. This further indicates that it would be more consistent and sensible to try out the setup of Trust Fund or FPF as constituted in the exemplary cases of the FIRD-TM (Nigeria) and the ICBG Project (Suriname), respectively, or try to gain from the experiences of the Kerala Kani Samudaya Kshema (KKSK) Trust (India). The setup of NBF and State Biodiversity Fund may be left as such as a guiding principle that the people can always opt to turn to if they fail in formulating their own contracts. Additionally, the guidelines on the subject of benefit sharing framed by the NBA should leave adequate space to formulate case-specific set of agreements as is aptly provided in Section 21.3 of the Act.

Lastly, India's Biological Diversity Act may be tested in the light of specific concerns of the users of natural resources and responses to these concerns as documented in the report of an international workshop submitted in the fourth meeting of COP to the CBD. One of the recommendations in this report was to recognize the need to lower the uncertainty surrounding ABS, to build confidence between providers and users and to maintain flexibility. These concerns and the responses are given in Table 5.2.

Table 5.2:
Responses to major policy concerns in ABS

Concerns of Users of Natural Resources	*Responses*
1. Users need a clear contact point and to know of institutions authorized to grant access to genetic resources in provider countries	Establish a focal point and competent authority(ies) with clear responsibilities in determining access applications
2. Users need legal certainty on access	Create a worldwide list of focal points and competent authority(ies) to be accessible through the clearing house mechanism
3. Users need clarity on ownership/tenure/related rights	Providers authorize relevant institutions to make full ABS arrangements (collection, scientific research and commercialization)
4. Need for corporate ABS policies by users	Elaboration of corporate ABS policies, through: promotion by governments; peer pressure, for example, through professional associations; voluntary policies/measures

Source: SCBD (1998b).

A closer look of the above four concerns shows that India's Biological Diversity Act of 2002 is an excellent tool with respect to the basic concept of benefit sharing and the utilization of biological resources. However, elaborate arrangements shall have to be made at local level, mostly in the BMCs and also at the level of SBBs, to make these bodies easily identifiable.

Agreement on TRIPS, the CBD and India's Biological Diversity Act: A Comparison

The Biological Diversity Act of 2002 came into force in India during the year 2003 after pretty long years of deliberations over its draft, and during all those years, the subject matters covered in the Agreement on TRIPS and the CBD were a hot topic. Therefore, the Act was always drawing attention of many environmentalists, NGOs, researchers and other professionals. It is not surprising that India's Biological Diversity Act of 2002 has been reviewed and evaluated by many authors on a number of wide-ranging aspects, including the aspects of benefit sharing and collaborative research projects. As regards the latter, Gadgil (2003) observes that the Act has the status of a complementary Act and will have to be used side by side with a range of other Acts, including, in particular, those pertaining to forest, wildlife, Panchayati Raj institutions, plant varieties and farmers' rights and patents. Another observation by Professor Gadgil that there is always a danger that the regulations may encourage harassment and corruption, rather than effective action, also raises concern. There are other possibilities of loopholes, as Gadgil observes further, that may render India's Biological Diversity Act toothless, for any biological resource that is considered a commodity, or biological material that is blended and mixed, may be exempted from the provisions of the Act.

If we analyse India's Biological Diversity Act in the light of existing examples of benefit sharing arrangements, the analysis may be wrapped up as shown in the following paragraphs:

1. Provisions have been made but careful detailing with respect to effective implementation is required with involvement of various stakeholders.

2. One of the major aspects requiring immediate attention is the Act's effectiveness in protecting the TK bases and equitable sharing of its benefits with on-site stakeholders.

3. The provision of requirement of prior intimation to the SBBs need to be smoothly executed so as not to be categorized as 'excessive regulation' as many reviewers apprehend, particularly when domestic companies are allowed free access to the biological resources for the sake of bioprospecting.

In a situation when India has already entered the new regime of patent legislation, and the utmost requirement is to measure up to the provisions in the Agreement on TRIPS and also to encourage our domestic innovators to go for more national and international patents, laws and regulations have to be conducive to and in harmony with the single main objective of equitable commercialization of our biological wealth. The alarm raised with the arrival of the WTO-initiated IPR regime as regards the protection of intellectual wealth of the TK and also the biological resources of the developing world should be looked upon as a great opportunity, for otherwise also the flow of biological wealth was continuing from the 'South' to the 'North', and the biopiracy was phenomenally occurring through various kinds of research programmes and collaborations. Now that this phenomenon is under focus, the debate continues about the equitable distribution of benefits from the biological resources, acknowledgement of the vital role of TK, innovations and practices, conservation of biodiversity, and also about the necessity of measures in the IPR legislation, such as norms for disclosure of the country and the community from which a patentable subject matter and information regarding its application are obtained, and also about the proof of the PIC.

Much of the relief regarding the vulnerability of the biological wealth of the 'South' is drawn from the CBD, and in India, as has been assessed in the foregoing section, the enactment of the Biological Diversity Act of 2002 is being viewed by many as a handy tool to regulate the ABS. In order to evaluate the comparative advantages and disadvantages in the three main instruments, the Agreement on TRIPS, the CBD and India's Biological Diversity Act of 2002, related to these issues, an analysis is attempted with regard to the following themes:

1. Aims and objectives
2. Enforcement of IPRs
3. TK
4. Access to genetic resources
5. Mechanism of benefit sharing
6. Access to and transfer of technology for facilitating research and training in developing countries
7. Special needs of developing countries, the exchange of information and international cooperation in science and technology
8. Financial mechanism
9. Dispute settlement and the penal clauses

The outcome of this analysis is shown in Table 5.3, and this analysis makes clear the pluses and minuses in the three important instruments of the IPR regime in India's context. After having analysed these three treaties and the enactment, the ABS mechanism may be discussed, keeping in view the strengths and weaknesses of these instruments.

While we discuss about the regulations for ABS accruing from the commercial use of biological diversity, we realize that the subject of protection of TK, the local innovations and cultural practices of the indigenous communities is the most complex issue. A suitable mode for protecting TK is still evolving in several tropical countries, and the most efficient measures that could achieve the desired results are still open-ended and flexible. The immediate action found suitable makes one inclined towards the registration of knowledge, develop a sui generis system for protecting TK and proceed with what India is undertaking presently. The most prominent attempts are (a) preparation of the PBRs and (b) the TKDL.

It is conclusively realized that the familiar knowledge with the people presents difficulties such as (a) the community holding the TK is not a legal entity; (b) the TK is held quite largely by individual healers across several communities, or communities of people parallel to each other, making it complex to decide the ownership of TK; and (c) the conditions of novelty and innovative step necessary for grant of a patent are not satisfied. An excellent approach is shown in the provisions of India's Biological Diversity Act of 2002 that bears great commonality with the biodiversity laws of many other developing countries. In continuation, the aspect of equitable commercialization

Table 5.3:

A comparison of the provisions made under the Agreement on TRIPS, the CBD and India's Biological Diversity Act of 2002

Theme	Agreement on TRIPS	Convention on Biological Diversity	India's Biological Diversity Act of 2002
1. Aims and objectives	To reduce distortions and impediments to international trade; to take into account the need to promote protection of IPRs; and to ensure that procedures to enforce IPRs do not themselves become barriers to trade (*Preamble*). Protection and enforcement of IPRs to contribute to the promotion of technological innovation and transfer of technology, to mutual advantage of producers and users of technological knowledge, in a manner conducive to social and economic welfare, to a balance of rights and obligations (*Article 7: Objectives*).	Conservation of biodiversity is common concern of humankind; States have sovereign rights over their own bioresources and are responsible for conserving and using their biodiversity in a sustainable manner; traditional dependence of indigenous and local communities on bioresources and the desirability of sharing equitably benefits arising from the use of TK are recognized (*Preamble*). Three main objectives are: (a) conservation of biological diversity, (b) sustainable use of its components and (c) fair and equitable sharing of the benefits arising out of the utilization of genetic resources (*Article 1: Objectives*).	To provide for (a) conservation of biological diversity, (b) sustainable use of its components and (c) fair and equitable sharing of the benefits arising out of the use of biological resources and knowledge (*Preamble*). [*Same as in the CBD.*]
2. Enforcement of IPRs	IPRs are private rights (*Preamble*); each member to accord to nationals of other members equal treatment of IPR protection as it accords to its own nationals (*Article 3.1*; further supported by *Article 4*). 'Patents shall be available for any inventions, whether products or processes' and 'patent rights shall be enjoyable without discrimination as	'The Contracting Parties, recognizing that patents and other IPRs may have an influence on the implementation of CBD, shall cooperate in this regard in order to ensure that such rights are supportive of and do not run counter to its objectives' (*Article 16.5*).	Permission required for IPR, in India or outside, for any invention based on research or information on a bioresource obtained from India (*Section 6.1*). NBA may impose benefit sharing fee or royalty or both or impose other conditions (*Section 6.2*). Applicants applying for patents, in India or outside, or for obtaining bioresource occurring in

			India or knowledge associated thereto, for research or commercial utilization or transfer of research results, shall obtain approval of NBA. Permission may be granted by NBA subject to some terms and conditions including royalty imposition (*Section 19*). NBA may take measures necessary to oppose the grant of IPRs in countries outside India on any bioresource obtained from India or knowledge associated thereto (*Section 18.4*).
	to the place of invention, the field of technology and whether products are imported or locally produced' (*Article 27.1*). *Article 27.3b* provides patent protection to (a) microorganisms, (b) non-biological and microbiological processes and (c) plant varieties. On infringement issues, burden of proof to lie on defendants (*Article 34*). Members to enforce action against infringements, along with actions deterrent to further infringements, 'applied in such manner as to avoid creation of barriers to trade' (*Article 41.1*).		
3. Traditional knowledge	*Agreement on TRIPS neither recognizes the contribution of TK nor contains any provision allowing a country's claim to enforce fair and equitable sharing of benefits from the patenting of its own genetic resources abroad.*	Knowledge, innovations and practices of indigenous and local communities to be respected, preserved and maintained, and promoted for wider application. Each CP to encourage equitable sharing of benefits arising from utilization of such knowledge, innovations and practices (*Article 8j*).	The GOI to endeavour to respect and protect knowledge of local people relating to biodiversity, as recommended by NBA through such measures, which may include registration of such knowledge at local, state or national levels, and other measures for protection, including sui generis system (*Section 36.5*).

(*Continued*)

Table 5.3:
(Continued)

Theme	Agreement on TRIPS	Convention on Biological Diversity	India's Biological Diversity Act of 2002
		Traditional cultural practices get emphasis with the provision for encouragement of customary use of bioresources in accordance with traditional cultural practices that are compatible with conservation or sustainable use requirements (*Article 10c*).	Every local body to constitute a BMC within its area for promoting conservation, sustainable use and documentation of biodiversity including preservation of habitats, conservation of microorganisms and chronicling of knowledge relating to biodiversity (*Section 41.1*).
4. Access to genetic resources	*No provision exists as regards the access to genetic resources of the Members.*	Each CP to endeavour to create conditions to facilitate access to genetic resources for environmentally sound uses by other CP and not to impose restrictions counter to CBD objectives. Resources provided are only those that are provided by CPs that are countries of origin or by CPs that have acquired them in accordance with the CBD. Access granted to be on MAT, subject to PIC of the CP providing such resources, unless otherwise determined. Each CP to carry out scientific research based on genetic resources provided by other CP with full participation of, and where possible in, such CP and to take measures to share results of such research works (*Article 15*).	Prior approval required for obtaining bioresource occurring in India or knowledge associated thereto for research or for commercial utilization or for biosurvey and bio-utilization; provision applicable to any non-Indian individual or corporate body having non-Indian participation (*Section 3*). Results of research relating to bioresources from India not to be transferred to non-Indian individual or corporate body having non-Indian participation without prior approval (*Section 4*). Restrictions not to apply to collaborative research projects if projects conform to policy guidelines and are approved by the GOI (*Section 5*).

| 5. Mechanism of benefit sharing | *No provision exists for benefit sharing.* | Dependence of indigenous and local communities embodying traditional lifestyles on biological resources recognized; desirability of sharing equitably benefits arising from the use of TK, innovations and practices relevant to biodiversity conservation (*Preamble*). Each CP to take measures to aim sharing in fair and equitable way, upon MAT, the results of R&D and the benefits arising from commercial use of genetic resources with the CP providing such resources (*Article 15.7*). These measures get support also from *Articles 16, 19, 20 and 21*. | NBA while granting approval may impose fee or royalty from commercial use of IPR (*Section 6.2*). No person (excluding local people and communities, growers and cultivators of biodiversity, and *vaidyas* and *hakims*) to obtain bio-resource for commercial use or bio-survey except after giving prior intimation to the SBB concerned (*Section 7*). Equitable sharing of benefits accruing from biological resources, their by-products, innovations and associated practices as per MAT and conditions with the local bodies and the benefit claimers concerned; benefit sharing through joint ownership of IPRs; transfer of technology; location of production or R&D units; monetary compensation and non-monetary benefits; setting up of 'venture capital fund' (*Section 21*). Benefit sharing money to be kept in NBF, to be used for channelling benefits to the benefit claimers; conservation of bioresources and socio-economic development (*Section 27.2*). State Biodiversity Fund to be used for: conservation of heritage sites; compensating the people economically affected by notification under *Section 37.1*; |

(*Continued*)

Table 5.3:
(Continued)

Theme	Agreement on TRIPS	Convention on Biological Diversity	India's Biological Diversity Act of 2002
			socio-economic development (*Section 32.2*). BMCs to levy charges by way of collecting fees for accessing bioresources for commercial use (*Section 41.3*).
6. Access to and transfer of technology for facilitating research and training in developing countries	Developed country members shall provide, on MAT, technical and financial cooperation in favour of developing and least developed country members. Cooperation to include assistance in the preparation of laws and regulations related to the enforcement of IPRs, prevention of their abuse and to include support regarding the establishment of domestic offices and agencies relevant to these matters and training of personnel (*Article 67*: Technical Cooperation).	CPs, taking into account the special needs of developing countries, shall promote and cooperate in the use of scientific advances in biodiversity research in developing methods for conservation and sustainable use of bioresources (*Article 12c*). Each CP to take appropriate measures so that the developing country CPs, which provide genetic resources, are provided access to and transfer of technology that makes use of those resources, on MAT, including technology protected by patents and other IPRs (*Article 16.3*). These measures get support also from *Articles 18 and 20*.	The GOI to develop national strategies, plans, programmes for conservation and promotion and sustainable use of biodiversity including measures for identification and monitoring of areas rich in bioresources, promotion of in situ and ex situ conservation of bioresources, incentives for research, training and public education to increase awareness with respect to biodiversity (*Section 36.1*). Access to biological resources to be approved by the NBA (*Sections 19 and 20*). The NBA also to determine manner for (a) transfer of technology, (b) location of production, R&D units in such areas that will facilitate better living standards to benefit claimers and (c) association of Indian scientists, benefit claimers and the local people with R&D in bioresources, biosurvey and bio-utilization (*Section 21.2*).

| 7. Special needs of developing countries, the exchange of information and international cooperation in science and technology | Without mentioning the term 'developing countries' here, the Agreement provides that members may, in formulating or amending their (IPR-related) laws and regulations, adopt measures necessary to protect public health and nutrition, and to promote the public interest in sectors of vital importance to their socio-economic and technological development (*Article 8.1*). | CPs to facilitate exchange of information relevant to conservation and sustainable use of biodiversity, special needs of developing countries accounted for, including the exchange of research results of scientific and socio-economic nature, specialized knowledge and TK (*Article 17*). CPs to promote scientific cooperation with other CPs, particularly developing countries, in implementing the CBD through development and implementation of national policies. Special attention to be given to strengthen national capabilities. Clearing house mechanism to be established to facilitate technical and scientific cooperation. CPs to cooperate for (a) the development and use of technologies, including indigenous and traditional technologies; (b) training of personnel and exchange of experts; and (c) establishment of joint research programmes and joint ventures for development of relevant technologies (*Article 18*). | Restrictions applicable to non-Indian individuals or corporate bodies having non-Indian participation under *Section 3* (prior approval for biodiversity-related activities) and *Section 4* (biodiversity-related research results not to be transferred) shall not apply to collaborative research projects involving transfer or exchange of biological resources or information relating thereto between institutions, including Government-sponsored institutions of India, and such institutions in other countries, if such projects conform to policy guidelines and are approved by the GOI (*Section 5*). |

(*Continued*)

Table 5.3:
(Continued)

Theme	Agreement on TRIPS	Convention on Biological Diversity	India's Biological Diversity Act of 2002
8. Financial mechanism	*No provision exists with regard to this theme.*	Detailed financial mechanism provided to enable developing countries to meet the agreed full incremental costs of implementing measures that fulfil the CBD obligations. New financial resources to be provided by developed country CP who may also provide financial resources related to implementation of the CBD through bilateral, regional or other channels. Mechanism to be on a grant or concession basis, to function under the guidance of, and be accountable to, the COP. Policy, strategy, programme priorities and eligibility criteria relating to the access to and utilization of such resources, to be determined by the COP (*Articles 20 and 21*). *Article 20.4* proffers that the developed country CPs ought to fulfil their commitments under the CBD related to finances and transfer of technology, taking fully into account that economic and social development and eradication of poverty are the overriding priorities of the developing country parties.	A three-tier institutional mechanism is provided to enforce the provisions: the NBF, the State Biodiversity Fund and the Local Biodiversity Fund at the national, state and local levels, respectively. The Act provides for grants or loans to these funds to be paid by the GOI or the respective state governments (*Sections 26, 31 and 42*). The royalties imposed for the commercial use of bioresources, the sums and other charges including the charges levied as fees for collecting bioresources and all other sums received as per the provisions shall be credited to these three funds depending upon their jurisdiction or the decisions taken for the purpose. The financial mechanism is well laid out and includes the activities for which the fund shall be utilized, the manner in which the funds will be channelled and the way the accounts shall be audited.

| 9. Dispute settlement and the penal clauses | The Agreement endorses the provisions of GATT 1994 for applying them to consultations and the settlement of disputes under this Agreement (*Article 64*). | CPs to seek solution initially by negotiation in the event of disputes between them concerning the interpretation or application of the CBD. If failed, third-party mediation to be requested. An unresolved issue may be taken up either for arbitration provided in the CBD or to the International Court of Justice, or may finally be submitted to a conciliation commission created for the dispute settlement (*Article 27*). | The offences under the Act are cognizable and non-bailable (*Section 58*). The penal clause (*Section 55*) has stringent measure for contravention or attempt of contravention of the Act, making it punishable with imprisonment up to five years, or with fine up to ₹10 lakh (even more to make it commensurate with the damage caused), or with both. Whoever obtains a bioresource for commercialization without giving prior intimation to the SBB or is involved in an activity that is restricted on the account of being detrimental to the conservation and sustainable use of biodiversity shall be punishable with imprisonment up to three years, or with fine up to ₹5 lakh, or with both. |

of biodiversity has been studied with the help of examples of six noteworthy arrangements for benefit sharing in Chapter 6. The study should help in working out a set of methodologies, if not exactly a model, for benefit sharing of the resources of a biodiversity-rich country in order to circumvent the implications of WTO, though the anomalies in the benefit sharing accruing from the biological resources were a happening even before the advent of the WTO.

6

Case Studies on the Commercial Use of Biodiversity: Analysis of the Benefit Sharing Arrangements

There are many examples of ABS arrangements that have been put into practice during the past 25 years, spanning continents and covering several countries. The great leap forward was initiated in Costa Rica where INBio (or National Institute of Biodiversity) came into being in 1989 and successfully launched an institutional framework to promote, administer and monitor bioprospecting with an aim to share the benefits accruing from biological wealth. These examples of the ABS arrangements are primarily based on carefully drafted contracts, including numerous multilateral contracts, between companies and biodiversity collecting institutions that have the potential to be extremely useful tools towards ensuring the equitable sharing of the short- and long-term benefits derived from the commercial uses of biodiversity. Most of these arrangements have an apparent flexibility to adapt to local conditions, beneficial to source countries in a way that they can be worked with even if a country lacks IP laws and ABS regulations.

The agreements under these arrangements were negotiated on a case-by-case basis, depending on the rationale behind particular bioprospecting enterprises. The benefit sharing arrangements in most of these examples focus on the contractual agreements, the signing amount paid and some non-monetary benefits. This makes them appear like a purely commercial venture, while they need to be discussed with a holistic approach, keeping the socio-economic aspects of indigenous and local communities in view. However, as Laird and Kate (1999) observe, very few contractual agreements are made public, and the details of commercial partnerships are often confidential. Also observed frequently is that the benefit sharing discussion totally circumvents thorny issues such as the balancing of benefits between the 'North' and the 'South' (GRAIN, 2000). These are the very issues that should be central to any benefit sharing discussion, but at the moment, they are largely absent.

Another startling aspect is that the details of commercial bioprospecting partnerships and other related contractual agreements might not be available in general because of the obligation to maintain top secrecy for such documents, yet there is no dearth of literature on aspects such as TK, biodiversity prospecting and conservation and bioprospecting partnership agreements, regulations and policies. The SCBD alone has published enormous literature on benefit sharing, and many others—the WTO, the WIPO, universities, NGOs and many of the bioprospectors—have come out with thousands of pages on various issues related to this subject. Many partnership agreements signed between the resource providers and the corporate institutions remain unpublished too, the most discussed accord being between INBio and Merck, which is a private contract and not open to public inspection (Lash, 1993).

As regards the policy and national regulations that should cover bioprospecting partnerships, it appears that many countries, which have ratified the CBD as well as the WTO, are still in the process of making necessary changes in existing laws or enacting new laws in accordance with the Agreement on TRIPS and more specifically drafting the regulations related to the access to genetic resources and benefit sharing as required in Article 15(7) of the CBD.[1] Furthermore, it is seen that many bioprospectors-turned-authors who have written about their benefit sharing experiences have adopted the pick-and-choose methodology. As a result, no corporate bioprospecting agreements—anywhere in the world—are currently public, and at the same time, the rhetoric around benefit sharing has become highly abstract and difficult to understand (GRAIN, 2000).

---

[1] Article 15(7) states that each CP shall take measures with the aim of sharing in a fair and equitable way, upon MAT, the results of R&D and the benefits arising from the commercial use of genetic resources with the CP providing such resources.

As an exception, it is observed by Laird and Kate (1999) that the models about which contract details are overtly available are the partnerships established under the ICBG programme and the arrangements established by other organizations such as the US National Cancer Institute (NCI). The ICBG programme is one of the cases related to access to and sharing of benefits arising out of the use of genetic resources, that were included by the SCBD in 1998 when the case studies were invited on different thematic areas. Overall, 15 case studies, as listed in Table 6.1, were invited: two cases from UNEP, two cases from governments and 11 studies from NGOs (SCBD, 1998a). A synthesis of these cases was brought out by the CBD in 1998.[2]

These case studies have been analysed, debated, embraced and yet criticized the world over. The most noted example so far in benefit sharing endeavours has been that of the INBio–Merck collaboration entered during 1991. The ICBG programme, included in the cases invited by the SCBD, which was taken up in 12 less developed countries, also is worth to be paid a very keen attention to. Few examples of benefit sharing cases relate also to the commercial use of endemic species. In order to elucidate, few interesting examples are the one from the Western Ghats of India, wherein the sharing of benefits of bioprospecting was initiated with the Kani tribe, and two examples from Africa: the San Hoodia and the argan oil cases. Another case related to fund accumulation for the purpose of benefit sharing is from Nigeria, where a trust fund (FIRD-TM) has been set up. A brief outline of some of these partnership programmes is given in the following sections to review the different benefit sharing mechanisms involved.

But before the various features of these classic cases are dissected, one thing that is very common must be stated. We notice that though found very promising, these cases have very little replication in reality. Why this is so is a matter of great concern. The present IPR laws in the international patent regime give protection to the knowledge generated by individuals, while the knowledge held traditionally by indigenous people in developing countries is exposed to frequent misappropriation. Nevertheless, the process is on as regards interpreting the provisions in the Agreement on TRIPS vis-à-vis the

[2] *Available on CBD website*: www.cbd.int/doc/meetings/cop/cop-04/information/cop-04-inf-07-en.pdf (Accessed in November 2014).

Table 6.1:
Case studies submitted for the fourth meeting of the COP to the CBD, May 1998

S. No.	Author	Title
Submitted by Governments		
1	Ministry of Environment and Forests, GOI	Benefit sharing model experimented by TBGRI (now renamed as JNTBGRI)
2	République du Mali, Ministre du développement rural et de l'eau	Programme test de gestion décentralisée de la pêche dans le Delta Central du Niger au Mali
Submitted by International Organizations and NGOs		
3	Aalbersberg, William G.: Korovulavula, Iso; Parks, Johne E.; Russell, Diane	The Role of a Fijian Community in a Bioprospecting Project
4	Anuradha, R.V.	Sharing with the Kanis: A Case Study from Kerala, India
5	Guerin-McManus, Marianne; Famolare, Lisa M.; Bowles, Ian A., Malone, Stanley A.J.; Mittermeier, Russel A.; and Rosenfeld, Amy B.	Bioprospecting in Practice: A Case Study of the Suriname ICBG Project and Benefits Sharing under CBD
6	Iwu, M and Sarah A. Laird	ICBG: Drug Development and Biodiversity Conservation in Africa: Case Study of a Benefit Sharing Plan
7	Laird, Sarah; Lisinge, Esterine	*Ancistrocladus korupensis*: A Species with Pharmaceutical Potential from Cameroon
8	Laird, Sarah; Lisinge, Esterine	Sustainable Harvesting of *Prunus africana* on Mount Cameroon: Benefit Sharing between PlanteCam Company and the Village of Mapanja
8a	Laird, Sarah; Lisinge, Esterine	Conclusion: The *Ancistrocladus korupensis* and *Prunus africana* Case Studies from Cameroon: Contrasting Benefit Sharing in the Pharmaceutical and Phytomedical Industries
9	Madzudzo, Elias	Communal Tenure, Motivational Dynamics and Sustainable Wildlife Management in Zimbabwe
10	Moran, Katy	Mechanisms for Benefit Sharing: Nigerian Case Study for CBD

S. No.	Author	Title
11	Rosenthal, Joshua P.	The ICBG Programme
12	ten Kate, Kerry and Amanda Collis	The Genetic Resources Recognition Fund of the University of California, Davis
13	ten Kate, Kerry, Laura Touche and Amanda Collis	Yellowstone National Park and the Diversa Corporation Inc.
14	ten Kate, Kerry and Adrian Wells	The Access and Benefit Sharing Policies of the United States National Cancer Institute: A Comparative Account of the Discovery and Development of the Drugs Calanolide and Topotecan
15	Vogel, Joseph Henry	Case Study no. 6: Bioprospecting: The Impossibility of a Successful Case Without a Cartel

Source: www.cbd.int/doc/meetings/cop/cop-04/information/cop-04-inf-07-en.pdf (Accessed in November 2014; previously accessed from: www.biodiv.org in January 2005).

interests of the biodiversity-rich countries, and meanwhile, several biodiversity prospecting contractual arrangements are undertaken the world over.

A review of six case studies relating to the different aspects of sharing of benefits accruing from the wealth of tropical biodiversity is covered in the following sections. Out of the six case studies reviewed here, the first four were picked up from the case studies submitted for the fourth meeting of the COP to the CBD held in May 1998 (see Table 6.1) and two are selected because of some uniqueness involved in them.

This review would help in conceptualizing a workable method for benefit sharing of the biological resources of tropical countries. A comparison between the provisions under the Agreement on TRIPS, the CBD and India's Biological Diversity Act of 2002 has been incorporated in the previous chapter as an exercise, keeping the interests of the biodiversity-rich tropical countries in the backdrop. It is evident from this comparison that India's Biological Diversity Act of 2002, which emerged as a representative of the ABS laws and regulations adopted (or in the process of being formulated) in several other developing countries, has a design quite in line with the objectives of the CBD and has a mechanism that may assist in protection of TK despite the strong mandate of the Agreement on TRIPS. It will be seen in the review of the case studies that the provisions of India's Biological Diversity Act are also quite in line with the approach adopted for access to the resources and the benefit sharing arrangements in the various partnership programmes of these cases. After trying to carve out the commonality in these cases and picking up the best of the provisions under India's Biological Diversity Act of 2002, a mechanism for access to the biological resources and an equitable sharing of benefits is presented in Chapter 7.

Case Study 1: The INBio-Merck Collaboration

This is the most talked about and classroom-discussed case study, in which the INBio of Costa Rica, a nonprofit, semi-public organization established by Costa Rica's MINAE, entered into the bioprospecting agreement with Merck & Co., a US pharmaceutical corporation, in 1991. Under the agreement, over a two-year period, Merck received 10,000 plant samples with information about their traditional use and agreed in its lieu to provide research funding of $1 million in two years and also to establish a processing laboratory at the University of Costa Rica. The INBio–Merck collaboration, known to be the first case of its kind, provides the scope of diverse activities an institution can perform, using a country's natural resources in a sustainable way. The collaboration initially looked like it was going to stay for decades. But it has lost its sheen now, and in fact, INBio entered into numerous other partnership programmes, yet what is believed today is that INBio is in great financial trouble (TWN, 2013; Wade, 2014). Well, whatever has happened has happened and yet the INBio–Merck collaboration stands tall with its uniqueness and still remains exemplary. The following description will show up the components of the collaboration that made this case the most renowned set of agreements: institutional strengthening, resource assessments and biodiversity inventory, drawing returns from biological wealth and financial sustainability and capacity building. Finally, over a commentary on this programme, it will be discussed what and where went wrong and what lessons could be learnt from this case.

It is interesting to note that INBio came into existence in 1989, and the collaboration with Merck & Co. set off in the year 1991, much ahead of the CBD that came in 1992 and was ratified in December 1993. A good many analysts belonging to various NGOs, researchers (including those associated with INBio) and other authors have studied and documented the landmark collaboration and the experiences subsequently gained in bioprospecting by INBio from different perspectives. Of course, the goal set for INBio was very ambitious, and the basic aim was to help conserve, study and use Costa Rica's biological diversity. Its founders, Dan Janzen and Rodrigo Gamez, coupled with elected president Oscar Arias promoted INBio and propagated a new awareness of the value of biodiversity in Costa Rica (Gibbons, 2003). Its relationship with the Government of Costa Rica was to have a mutually supporting role. While INBio needed the government's help in seeking international support and maintaining the conservation areas as its principal work centres, the Government expected the support of INBio in biodiversity prospecting, information management, planning and policy making and in representing the country at the international environmental forums.

The agreement leading to the INBio–Merck collaboration deals with each party's obligations, exclusivity of arrangements, confidentiality of information, inventions and patents, payments in terms of royalties, indemnities and specific clauses to the effective length of the agreement and its termination. Under the agreement, INBio granted Merck the right to evaluate the commercial prospects of a limited number of plant, insect and microbial samples collected in Costa Rica's 11 conservation areas. INBio also agreed to establish the facilities for collection and processing of plants, insects and environmental samples, and also to hire and train staff for the purpose and to provide training in Merck facilities. Merck in turn agreed to evaluate the samples for potential activity as health and agricultural compounds. A system was agreed to for identifying products, which could earn royalty. As a part of one of the agreements, MINAE was to receive 10 per cent of Merck's prospecting fee out of INBio's share. Merck agreed to pay the royalty, to be shared equally between INBio and MINAE, on the profits of any future pharmaceutical product or agricultural compound that would be isolated or developed from an INBio sample.

According to an INBio spokesperson, 2 per cent royalty on 20 products could yield more money than the Costa Rican government got from exports (TED, undated). Nevertheless, this is no indication of any predetermined rate of royalty share, as the percentage of royalty under the agreement was considered as secret, and it still remains a secret. It is believed that this percentage might have been within the range of usually accepted percentages for agreements of this kind, and it could be between 1 per cent and 3 per cent for all the products developed through the agreement. The INBio and Merck had the following obligations in their agreement (UNCTAD, 2000).

INBio's Obligations

1. INBio will establish the necessary facilities in Costa Rica for the collection and process of plants, insects and environmental samples.
2. INBio will hire and train the necessary personnel for the collection and process of the samples. Merck agrees to give training in its laboratories to INBio's personnel or to whom INBio appoints.
3. INBio will provide Merck with a specific number of plants, insects and environmental samples on yearly basis for a period of two years as established in the work programme of the Agreement.
4. The samples of plants and insects will be processed in a laboratory established by INBio at the University of Costa Rica through a subcontract of services and at INBio.

Merck's Obligations

1. Merck will provide INBio with a research fund of $1 million during the first two years of the agreement and will contribute with the laboratory equipment and the required materials for INBio to operate the laboratories for processing the samples at INBio and at the University of Costa Rica.
2. Merck will assess the samples supplied by INBio through biological experiments owned by Merck to detect potential activity of compounds for use on human and animal health and for agriculture.
3. Merck will notify INBio of any activity capable of reproduction identified in the samples supplied by INBio.

4. Merck will give a unique numeric identification to all the samples sent by INBio and will keep an identification system that will allow the two parties to identify all the products from which there is a possibility to obtain royalties under the agreement.

The Benefit Sharing Arrangements

The collaboration specifies that 10 per cent of the initial $1 million and 50 per cent of any royalty will be invested in biodiversity conservation through MINAE. The remainder of the royalties will be used to preserve the environment at the discretion of INBio's Board of Directors. The exclusivity clauses of the agreement mention about restrictions such as non-supply of samples to other companies, or rejection of the collection of a sample if it is impossible to get the sample for logistic or biological reasons. The agreement also has clause on the confidentiality issues, providing that during the agreement and for a period of seven years after the agreement is concluded, the parties agree not to reveal to third parties confidential information obtained during the period of the agreement. It also has the provision that each party could publish the research results after providing the other party the opportunity to examine the publication.

It also has the provision that the inventions created within the investigation will belong to Merck, and Merck will be responsible for requiring and registering in a proper manner the applications for the award of the patents. About the payment of royalties, the agreement has in it the method that the royalties will also apply to any product derived from or analogous to these compounds, to chemical compounds derived from living microorganisms isolated from environmental samples or from samples of dead tissue. The criteria and terms of the Research Collaborative Agreements used by INBio have been summarized by Rodrigo Gamez (2003), one of the founders of INBio, as they appear below:

1. Access is limited to a given amount of samples and is facilitated for a limited period of time, under the terms established by national legislation and a framework legal agreement between INBio and MINAE;
2. Technology transfer and capacity building must be ensured, and based on the existing capabilities, major part of research

should be carried out locally, with costs covered entirely by the industrial partner;

3. The upfront payment of a minimum of 10 per cent is included in the research budget, and is transferred directly to MINAE for exclusive conservation purposes; and

4. The discovery and development of a product is restricted to non-destructive uses of natural resources and must be entirely consistent with the national legislation dealing with access to genetic resources.

Gamez (2003) states that the benefit sharing mechanisms should include (a) milestone payments, and (b) percentage in royalties, both to be shared on a fifty-fifty basis on net sales of the final product along with joint patents and publications. According to the collaborative agreement with MINAE, INBio conducts its bioprospecting activities, with a few exceptional cases, only in MINAE's protected areas. Contrary to the situation prevalent in many countries throughout the world, protected wildlands in Costa Rica have no inhabitants, local farmers or indigenous people. This is the reason why the distribution of monetary benefits in the INBio–MINAE agreement does not contemplate these particular sectors of society. Termination clause, as well as the clause for sub-licensing of the contract under certain conditions, was also provided under the agreement.

The collaboration was entered in November 1991, and was subsequently renewed twice (July 1994 and August 1996). The main programmes taken up under the INBio–Merck collaboration are summarized in the following paragraphs.

National Biodiversity Inventory

One basic component was to provide training to local parataxonomists who were the group of men and women from the rural areas of Costa Rica for collecting and mounting plant, insect, mollusc and fungi specimens. They also collected field information associated with these resources. Apart from supplying information on the natural history of the specimen, these individuals acted as disseminators in their communities and were active participants in the joint processes between INBio and the National System of Conservation

Areas. The trained people would work at the centres distributed across the country and would have the responsibility of categorizing the selected materials and teach the local communities about their environment.

This programme also sought to create an inventory of the wild-life taxa. INBio had the responsibility for providing necessary training, planning and management. On the reports of the paratax-onomists, the process of registration was to begin at INBio, in steps, into the National Biodiversity Inventory, the National Biodiversity Information Management System and finally the international net-work of taxonomists and collections. The inventory constituted the basis upon which INBio would perform the rest of its activities, and the Government was to use this information in its decision-making process. Additionally, various institutions involved in the programme would directly benefit from this inventory.

Biodiversity Information Dissemination Programme

This programme was designed to promote a new level of biological literacy, a knowledge that was perceived to be lost, and to strengthen the process of evolving a society whose ethical values have respect for nature and wise management of its resources. To be precise, INBio included the following actions: supplying historic and taxonomic information to educational centres, providing advice in commercial development of conserved areas, participating in policy making and national or international planning gatherings, training personnel and producing biodiversity literature.

Biodiversity Information Management Programme

The main goal here was to build capacity for efficiently managing the information gathered in the inventory programme and provide it further, in an appropriate interface form, to the potential users. The amount of information and its complexity was enormous, help-ing INBio in establishing partnerships with American corporations and other conservation institutions aimed at computerizing and net-working.

Biodiversity Prospecting Programme

The main goal of this programme was to find new compounds and chemicals from the isolates and other material collected from the flora, insects and microorganisms inhabiting the protected areas, and use the results of research in the pharmaceutical, medicinal, biotechnological and agricultural sectors. As part of its effort, INBio developed a Biodiversity Prospecting Unit that engaged in the collection and marketing of biodiversity samples and collaborated with local and international institutions in conducting additional processing of samples and initial research into the chemical and biological properties of Costa Rican biodiversity (Aylward, 1995). The programme included a 10 per cent minimum collaboration to MINAE from each research project to sustain its conservation activities, technology transfer, training and compensation for services and information supplies. The research budget also supported the national science and processing infrastructure.

The monetary and non-monetary benefits derived by INBio from bioprospecting have been enlisted by Gamez (2003) as shown below:

Monetary Benefits:
1. Totally funded local research budgets
2. Technology transfer and infrastructure
3. Upfront payments for conservation
4. Strengthening of research capacity of local scientific institutions
5. Milestone and royalty payments shared with MINAE

Non-monetary Benefits:
1. Training and empowerment of human resource
2. Technology transfer
3. Shared research results and information
4. Negotiations expertise developed
5. Market information
6. Improvement of local legislation on conservation issues

According to Laird and Lisinge (2002), the INBio–Merck collaboration has been very attractive figuratively: of the $1,135,000 advance payment made to INBio by Merck in 1991, for example, $100,000 was provided to MINAE, Costa Rica. The biodiversity prospecting agreements yielded more than $390,000 to MINAE; $10,000 to conservation areas; $710,000 to public universities; and $740,000 to cover activities within INBio, primarily the National Biodiversity Inventory. INBio also signed an agreement with the Costa Rican Ministry of Natural Resources, Minerals and Energy (MIRENEM) that set forth the cooperation terms that included the terms for specimen collection for research, as well as for chemical prospecting. Aylward (1995) observes that this agreement formalized a few of the practices regarding the transfer of benefits from INBio to the national parks, and the contract of returning part royalty to the body responsible for biodiversity protection (with an arrangement to split royalties with the national parks) was novel in this respect.

The INBio–Merck partnership persisted for about eight years, and the turnaround of the landmark collaboration was an eventuality in the year 1999 when Merck ended the agreement in order to concentrate on analysing the collected samples. But INBio learnt big lesson of negotiation and benefit sharing agreements during its unyielding and yet poised course of action. As a result, INBio could sign around 11 agreements of this nature (UNCTAD, 2000), all of which embraced the following seven basic aspects:

1. Direct payments in cash or knowledge exchanges (equipment, training, technological know-how).
2. Payment of a significant percentage of the initial budget of the project (10 per cent) and the returns of the commercial uses of the products (50 per cent).
3. Cooperation clauses that stipulate the gradual translation of the investigation processes to the supplier country, in order to create new jobs and achieve industrial development.
4. Minimum exclusivity.
5. Agreement on the samples property and patents property.
6. The use of chemistry synthesis, semi-synthesis and domestication of the living sources, in order to avoid the continuous extraction of the biotic material.
7. Legal mechanisms that will provide protection to both parties.

In all these years, INBio had been working in partnership with scientific research centres, universities and pharmaceutical, biotechnological, agro-industrial and cosmetic industries that were mutually beneficial to all parties. These pioneering agreements provided significant returns to Costa Rica while simultaneously assigning economic value to the natural resources. INBio truly created numerous bioprospecting partnerships, and as Laird and Kate (2002) find, INBio entered into commercial partnerships with Bristol-Myers Squibb (B-MS—US pharmaceutical company), Phytera (US biotech company), INDENA (Italian manufacturer of botanical extracts and phytochemicals for the pharmaceutical, botanical medicine, cosmetic, food and other industries), Givaudan Roure (Swiss–US company interested in new fragrances) and Analyticon (German service and contract research company). INBio also entered into contracts with Diversa (USA), La Pacifica (Costa Rica) and the British Technology Group (BTG) (Laird and Lisinge, 2002).

INBio is much appreciated example for one more reason. It has developed its own research capacity so that it can investigate diseases (often ignored by transnational corporations) and local agricultural problems. For its pioneering work, INBio has earned international admiration and has won many prizes that recognize it as an institution unique in showing how science can be used for the benefit of humankind. The inventory activities and participatory approach, which included students, women and rural inhabitants, received the greatest recognition.

Other Lateral Collaborative Agreements

Let us look at some other mutually beneficial alliances extended by INBio with many other scientific research centres, universities and industries. In addition to channelling funds to support the country's Conservation Area Development Schemes, these agreements provide sizeable returns to Costa Rica through the process of assigning economic value to its genetic resources. On the lines of its collaboration with Merck, other examples of some of the benefit sharing agreements signed by INBio with companies from different industries (Laird and Kate, 2002; Laird and Lisinge, 2002; UNCTAD, 2000) are as follows:

1. *INBio–Givaudan Roure Agreement.* This agreement with the Swiss–US company interested in new fragrances was signed in 1995 with the objective to explore the potential of the biodiversity fragrances and aromas, which could be eventually synthetically reproduced.
2. *INBio–DIVERSA Agreement.* This agreement with the US biotechnological company was signed in 1998 with the objective to explore new enzymes in aquatic and terrestrial microorganisms from Costa Rica.
3. *INBio–INDENA Agreement.* This agreement with the Italian manufacturer of botanical extracts and phytochemicals was signed in 1996 with the objective to obtain anti-microbial potential compounds, which could be used as active ingredients for the pharmaceutical, botanical medicine, cosmetics, food and other industries.
4. *INBio–BTG Agreement.* This agreement was signed in 1992, and its main objective was the investigation, characterization and production of a product with nematic activity contained in a tree from the dry Costa Rica's forest.

Above and beyond the ties with industrial sector, INBio has signed agreements with academic, nongovernment and government sectors also with almost similar objectives of sharing its resources and knowledge in order to achieve solutions to specific problems. Two significant cases out of these have been with (a) the University of Massachusetts, signed in 1995 with the objective of locating in the environment the components that contain insecticide activity; and (b) the University of Strathclyde, with the objective of supplying plant samples to various Japanese industries (UNCTAD, 2000).

The agreements with BTG and Kew Gardens, and with B-MS and Cornell University, have schemed for 10 per cent of research fund to be directed to national parks system. INBio also received funds from a variety of US-based private foundations, international environmental NGOs and bilateral assistance agencies. Contributions from these agencies, as in case of other agreements, also included funds earmarked for national parks, human capital development and technology transfer and inventory activities.

Commentary on the Collaboration

The InBio–Merck collaboration generated extensive interest among researchers, planners, corporate world, public institutions and several NGOs in different ways. Many find the contract appearing to be an innovative and figuratively attractive collaboration. And yet a number of NGOs and others are critical about the agreement. A major shortcoming of the collaboration is reported to be that many aspects of the initial 1991 agreement and subsequent contracts are secret. Carolyn Crook, who was granted an International Development Research Centre (IDRC) Doctoral Research Award to assess the current and future benefits of bioprospecting agreements and the relative distribution of benefits within Costa Rica and Peru, states (Eberlee, 2000) that royalty rates are held strictly confidential, partly so that INBio may be able to negotiate higher royalty rates in future agreements with other companies who, if they know the 'going' rate, would resist paying more. She adds that quantifying the extent and distribution of benefits is important before unduly relying on these agreements as a mechanism for promoting conservation. Still, irrespective of this fact, we can learn great lessons in bioprospecting partnership deals through this analysis, and it would be pertinent to craft and outline a scrupulous commentary on the INBio–Merck collaboration.

Mentioning that Merck enjoys the exclusive right to draw patents while the percentage of royalty accruing from the products derived from the explorations is being kept confidential, Sharma (2004) observes that INBio provides unrestricted access to 'scout the tropical forests for a paltry fee' and strongly argues that if this collaboration is being construed as a successful benefit sharing experiment, it will be useful to know what will constitute an exploitative collaborative effort. In the context of the INBio's agreements on the principle of splitting royalties with the national parks, Aylward (1995) states that at the present time, all such royalty agreements and distribution mechanisms remain untested. Campbell (2002) states that the upfront payment of $1 million by Merck was a small fee for a company that has annual profits exceeding $8 billion. Gibbons (2003) affirms that Merck was drawn to Costa Rica's rainforests due to the rumours of 'green gold' of the country.

Many authors (Campbell, 2002; Hurlbut, 1994; Mott, 1993; Nygren, 1998) believe that bioprospecting deals can be seen as one more way for the technology-rich 'North' to buy the resources of the biodiversity-rich 'South' at a relatively low cost, for the developing countries often cannot afford the pharmaceuticals developed from the crude genetic resource they supply, thus further worsening the unequal terms of trade. Campbell (2002) argues that the IPR to resultant products are ceded to Merck, and therefore lost to the parataxonomists. Another criticism against INBio has been that the indigenous population was not properly consulted, and signing of the deal was behind closed doors. As was observed by Carolyn Crook, the indigenous communities have not yet shared in the economic benefits to any great extent, though simultaneously there is an encouraging statement that Costa Rica is earning revenue from resources for which it previously received nothing (Eberlee, 2000).

INBio's initiative and experience have been sharply criticized even on the ground that it was an advanced form of biopiracy. Nevertheless, the commercial use of biodiversity is more positive approach, as Gibbons (2003) quotes Janzen and Gamez saying that if the biodiversity is not made a commodity and if it is not competitive with other commodities to outperform other uses of land, it will be lost. Another commonly heard phrase today is: 'No patents, no benefits!' This indicates that the biodiversity-rich 'South' is now ready to adopt tough attitude at the international negotiations. It is therefore well understood why the lateral collaborations entered into by the INBio have been upheld as commendable venture by many agencies. Professionals who have known the INBio–Merck agreement well assert that Costa Rica is uniquely positioned to take advantage of the collaboration, and yet other pharmaceutical companies and other developing countries including Mexico, Indonesia and Kenya are studying the INBio–Merck agreement as a preliminary step to setting up their own deals (Mulder and Coppolillo, 2005; Reid, 1993; TED, undated).

The collaboration with Merck deserves to be recorded as the one that comprised textbook type of multilateral agreements that INBio entered into, a process that was going on while it was simultaneously mentioned in some corners that financially INBio was in doldrums. According to an article by Edward Hammond, TWN's research consultant on biopiracy, INBio is in the midst of a deep financial and organizational crisis, and the Institute cannot pay utility bills and other maintenance costs for its sample collections, which consist of more than 3.5 million samples of the country's biodiversity (TWN, 2013). Perceptively, there are two things that went wrong with INBio. One is that it created a collection of more than 3.5 million samples estimated to cover nearly one-third of Costa Rica's biodiversity, and then it was unable to bear the cost of maintaining the huge collection (TWN, 2013), which was not paying back in any way. Another blunder was to create the INBioparque, a biodiversity theme park spread over 7.2 hectares that aimed to attract tourists, after borrowing $7 million, which again was a financial failure (Wade, 2014). The collection will now pass on to the government-owned National Natural History Museum, and the INBioparque also has been purchased by the government (TWN, 2015) in order to rescue INBio from the financial trouble. INBio seemingly is trying now to transform itself to achieve a more stable future (Wade, 2014).

The pioneering collaboration may be termed as a story that ends, but this story also left a legacy of the most diversified multilateral collaborations and multidimensional endeavours in the field of commercial use of biodiversity. The future planners of biodiversity conservation, research and business may benefit a lot from the huge information generated by this case and the financial intricacies in which INBio was trapped.

Case Study 2: Suriname ICBG Project

The Suriname ICBG Project, initiated in 1993, was one of the initial projects of the ICBG programme. The umbrella ICBG programme, launched in 1992, was the product of a workshop in 1991 sponsored by three agencies of the US government: the NIH, the NSF and the US Agency for International Development (USAID). The NIH emerged as the administrator of the programme and announced a competition for large grants for research into the pharmaceutical potential of international biodiversity (Greene, 2004). The programme involves diverse public and private institutions including universities, environmental organizations and pharmaceutical companies, collaborating on multidisciplinary projects with focus on the potential relation-ships between drug development, biological diversity and economic growth. The size of the umbrella programme is indicated by the size of the awards (approximately $600,000 per year[3]) for the seven ongo-ing projects taken up currently in seven countries.

The ICBG represents a novel experimental programme that is one of the first large-scale attempts to design such a multidisciplinary approach to drug discovery and to gather evidence on the feasibil-ity of bioprospecting as a tool for conservation and economic devel-opment (Rosenthal, 1997b). Several projects under the ICBG, taken up in Jamaica, Samoa, Tonga, Argentina, Chile, Mexico, Cameroon, Nigeria, Peru, Vietnam, Laos, Uzbekistan, Kyrgyzstan, Tajikistan and Kazakhstan are already complete, while some projects are still in pro-gress, in some of which even a second phase has been launched. The projects are in progress in Costa Rica, Fiji, Indonesia, Madagascar, Panama, Papua New Guinea and Philippines.

The constituents of the Suriname ICBG project, which is already complete, make a very interesting set of examples as they include a novel design for active participation of various partners of the programme, meticulous procedures for drafting of contractual agreements, equita-ble sharing of benefits, capacity building and a unique fund manage-ment plan. To conclude the case, a reference shall be made to two other ICBG programmes—the Peruvian and Mexican programmes, the first one because of a clue to great drug discovery prospects and the second one for its failure due to strange local conflicts.

The Suriname ICBG project was tailored to the particular condi-tions of Suriname and of the Maroon communities involved, making an effort of its own kind that addresses three issues: drug discovery, biodiversity conservation and sustainable economic growth. The project links the three issues mentioned into a multi-country con-servation and development effort funded and guided cooperatively by the NIH, NSF and USAID. The cooperating NIH components are the Fogarty International Center (FIC), NCI, National Institute of Allergy and Infectious Diseases (NIAID), National Institute of Mental Health (NIMH), National Institute on Drug Abuse (NIDA) and the National Heart, Lung, and Blood Institute (NHLBI). The pro-gramme envisaged that benefit sharing might provide clear incentives

[3] *Source*: www.icbg.org (Accessed in May 2014).

for conservation and sustainable use of biodiversity. It was based on the belief that the discovery and development of pharmaceutical and other useful agents from natural products could, under appropriate circumstances, promote scientific capacity development and economic incentives to conserve the biological resources from which these products were derived. The programme is designed to provide models for sustainable use of biodiversity, and to explore the viability of bioprospecting as a means to:

1. Improve human health through discovery of new drugs from natural products;
2. Conserve biodiversity through local capacity building (training, infrastructure development, etc.) to manage natural resources; and
3. Promote sustainable economic activity of communities, primarily in the less developed countries that are the source countries of natural resource.

The project offered useful working model for national and international policy discussions related to biodiversity conservation incentive measures, technology transfer, IP and benefit sharing. In fact, each of the ICBG projects is run as multidisciplinary, multi-institutional programme by an academic principal investigator, who directs his or her own research programme in natural products chemistry, drug development or ethnobotany, and coordinates the activities of several associate programmes that include other academic research institutions, NGOs and, in most cases, a commercial pharmaceutical partner. The programme has certain uniqueness as primarily it seems to encompass all the basic tenets of an equitable sharing of benefits. The awards are in the form of cooperative agreements, rather than grants. Furthermore, formal agreements governing the treatment of IP and benefit sharing are required of all applicants prior to making an award. The funding agencies are not parties to the research and benefit sharing agreements, and the principles require that full disclosure and informed consent are carried out, that both near- and long-term benefits are shared with appropriate source country communities and organizations, that local laws and customs are respected, and that credit is given to local indigenous or other intellectual contributors wherever possible (Rosenthal, 1997a).

The Suriname ICBG project also adhered to an experimental and largely an inclusive approach to drug discovery. Appreciating that royalties from new drug development were distant possibilities and therefore they must not be the only economic incentive, the capacity-building approach was seen as a more immediate and sustainable economic development process. Therefore, the programme design included (Rosenthal, 1997a):

1. Active participation of host country individuals and organizations from the planning stage onwards;
2. Multidisciplinary research on diseases of both local and international significance;
3. Local training and infrastructure development in both drug discovery and biodiversity management;
4. Biodiversity inventory and monitoring; and
5. Equitable IP and benefit sharing arrangements.

The project design also included an appropriate plan for the indigenous people, including traditional healers, to pay them for their services and training in collection and identification techniques. As regards the flow of various funds and the accounting of the payments for the services and other benefits, a fund called Forest People's Fund (FPF) was articulated. This was masterminded to ensure flowing back of both near- and long-term benefits to the collaborating communities, even if the ethnobotanical knowledge was not used in the research process. The functioning of FPF has been included later in this chapter. The implementation of the project was pursued through a signed contract and a statement of understanding. The project participants from universities, botanical gardens and pharmaceutical companies were to carry out botanical and ethnobotanical collections and inventory, extraction, screening, chemistry and drug development with involvement of local people. The process is elaborated in the following section.

The pharmaceutical partner involved in the project was the B-MS, and four other institutions were involved, each identified to carry out a specific role—another uniqueness of the arrangement. These institutions were: the Virginia Polytechnic Institute and State University (VPISU—a state-funded US university), Conservation International (CI—an NGO), Missouri Botanical Garden (MBG—a US botanical research institution) and Bedrijf Geneesmiddelen Voorziening Suriname (BGVS—pharmaceutical company of Surinamese government). The respective roles of these institutions were to carry out the screening and isolation studies on extracts of interest to them, to focus on conservation and economic development work as well as plant collection, to carry out plant collection work and, lastly, to play key roles in coordinating the in-country work and preparing extracts for study (Cao and Kingston, 2009).

Preparation of Contracts and Benefit Sharing Arrangements

For structuring of the contracts, the ICBG projects begin with a type of contract (a cooperative agreement) between the US government and a principal investigator at a US university. The government agrees to fund the project as described in the application, contingent upon the fulfilment of the set of principles. This exercise is run parallel to the identification of different partners. The CI took the lead in developing the Suriname ICBG proposal and, early in 1994, executed the 'Letter of Intent' with the Saramaka Maroons.[4] This was followed in other ICBG programmes, and it was effectuated that each party to the agreement had independent legal advice during the negotiation process. Normally, the negotiators do not have adequate legal or commercial experience to negotiate agreements without competent legal counsel, and industrial partners frequently get advantage under such situations (Rosenthal, 1998). For the ICBG, the programme leaders lay out the basic principles of their agreements and, with the aid of an attorney, develop the first draft. Subsequently, each party sends the draft to their legal advisors for careful analysis to ensure that their needs are being met and potential confusion and shortcomings are avoided. The funding agencies too, although not included as party to the research and benefit sharing arrangements, are asked to develop workable agreements within the general framework of these basic principles only to fit the nature of the organizations, countries, communities and resources involved (Rosenthal, 1997b).

---

[4] Saramaka Maroons are the people of African descent whose ancestors had moved to the South American country, the Republic of Suriname, during seventeenth and eighteenth centuries to work in sugar, timber and coffee plantations. Maroon is used for the group of people who could escape slavery, and out of the six Maroon tribes, Saramaka people form the largest group in Suriname.

The ICBG contract also guarantees confidentiality so as to protect the potential profits of all parties should a drug be developed. The parties may not share information, results and data from the project for five years after the contract terminates, unless the recipient of the information already has legal access to it. The project also promises even-handed sharing of benefits accruing from the components of the project. The long-term research agreement that controls the ownership, licensing and royalty fee structure for any potential drug development accomplishes an equitable benefit sharing. The 'Statement of Understanding' between the Granman, CI-Suriname and BGVS that further defines the parties' intentions regarding the distribution of royalties among Surinamese institutions, the transfer of technology and other forms of non-monetary compensation to Suriname and other tools also accentuate equitable benefit sharing. Rosenthal (1998) enlists the following four types of benefits in the ICBG programme:

1. *Royalties*: A percentage of earnings from commercial sales by the licensing partner may be agreed upon in the initial agreement, or the agreement can specify a range and require the parties to negotiate the final rate on a case-by-case basis. Some issues to consider in royalty structures include: (a) relative contribution of partners to invention and development, (b) information provided with samples and (c) novelty or rarity of sample organisms.
2. *Advance payments*: Access fees may take the form of lump sum or milestone payments, per sample fees, payment for resupply of samples or in-kind contributions of equipment, training, medicines, etc.
3. *Equipment, training and infrastructure*: Commercial partners or nonprofit funding organizations may provide resources to help build the capacity of source country partners to execute current or future needs for bioprospecting research, medical care, biodiversity management and so on.
4. *Priority research areas*: Agreements can require that locally important, but understudied, diseases and indigenous therapies will be investigated by commercial and other scientific partners. Additionally, they can focus specimen collections and identification on geographical areas or biological groups that are high priorities for conservation needs.

Under the contractual agreement of the project, decisions regarding inventorship and ownership are made according to the US patent law. The ICBG also provides for joint ownership of patents where the product is a result of collaborative work with the local healer. B-MS, the pharmaceutical partner, agreed to pay all costs of patenting and patent protection where it was to be one of the joint partners of the patent (Guerin-McManus et al., 1998). The project provided that the terms of the licence, including fees and specific royalties, must be negotiated in good faith between B-MS and the owners of the invention. Licensing rights within the ICBG agreements were to be allocated to all inventors, and if there was any joint ownership of an invention between a local healer and B-MS, then both parties were supposed to negotiate the licensing provisions and control the use of the invention.

The Suriname ICBG project also offered that if any products were commercialized from ethnobotanical collections, an established benefit sharing plan prescribes (Guerin-McManus et al., 1998; Rosenthal, 1997a) that 50 per cent of Suriname's share of any future royalties will go to the FPF and the other 50 per cent will go to various ICBG partners such as the government-owned pharmaceutical company, the University of Suriname Herbarium, the CI-Suriname (10 per cent to each), Suriname Forest Service and Stichting Natuurbehoud Suriname (5 per cent to each) and the rest 10 per cent to be reserved for future collaborating institutions. On the other hand, if a drug is derived from random collections, the FPF's share is reduced to 30 per cent, while 70 per cent goes to other ICBG partners—10 per cent to each and 20 per cent to be reserved for future collaborating institutions.

Forest People's Fund

As part of an ICBG project, CI developed the FPF, a mechanism through which upfront compensation (the start-up money) and future royalties from new drugs developed by the pharmaceutical partner (B-MS) could be returned to the members of the indigenous communities of Suriname. The FPF was established in 1994 with a $60,000 total advance payment from B-MS, with additional annual contributions of $20,000 (Moran, 2000). The FPF Foundation administered the fund according to the Foundation's by-laws.

Conceptually, the FPF provides a concrete example of how local people who share their own knowledge can also share the benefits that accrue from that knowledge. It was agreed that the funds would be used for projects involving community development, biodiversity conservation and health care. The by-laws of the fund were written by Surinamese participants and were governed by the laws of Suriname. It was provided in the by-laws that the fund would be managed by a well-represented five-member Board of Directors consisting of two members from indigenous communities, two from CI-Suriname and one from the Department of the Interior. The Board also reviewed the proposals for the allocation of advance payment and future royalties. In addition, the FPF was operated to create conservation incentives, to finance sustainable management projects, to provide research and technology exchanges and also to support other projects involving community development, biodiversity conservation and health care.

It should be appreciated that the idea of returning benefits to indigenous and local communities is in general quite complicated in practice. Frequently, the funds are focused on a range of components and community issues such as improving health services and sustainable use of local resources as mentioned above, and the benefit sharing plan also includes contingent benefits to several government agencies (e.g., the National Herbarium), for the Government organizations that manage natural resources are generally underfunded. Still the mechanism of fund management, as in the FPF, which is outspread on a well-laid-out foundation, is arguably the best practice under complicated state of affairs.

Commentary on the Project

Edward C. Green, an independent US consultant, along with others, conducted an evaluation of the Suriname ICBG project in 1997. The team conducted interviews with the ICBG source country partners, with other relevant organizations and institutions and with local leaders, traditional healers and members of the village communities.

It was found in this evaluation (Green et al., 1999) that the Saramakas were the most traditional of the six Maroon societies, and their traditional ethnomedicinal knowledge probably was more extensive. Therefore, they only were found participating in the project at the time of evaluation. The funds had also been used to purchase outboard motors and develop a canoe taxi service to transport people and goods upriver to Saramaka territory, to train local people in catering, with ecotourism as target, to purchase sewing machines and training women, to supply tools and training for commercial gardening projects, woodcarving, carpentry and many other things. This is a fair example of acting locally even under an international project.

In the context of its uniqueness and the meticulous procedures adopted for drafting of contractual agreements, Rosenthal (1997b) mentioned that the goal to promote sustainable economic activity was accomplished through benefit sharing agreements that used 'novel contractual agreements'. The agreements between the ICBG Suriname and the FPF, however, did not provide specifically for the sharing of royalties with traditional healers. For the most part, the ethnobotanical information was made available to the ICBG project by traditional healers, and in the context of providing some kind of special compensation to them, one factor under consideration was that it challenged Maroon tradition for healers to sell knowledge about medicines to outsiders, even though their paramount chief and local chiefs voted in favour of participation. As ethnobotanical collecting began to develop, traditional healers expressed a preference for being paid about five dollars a day for their participation. Green et al. (1999) mention that given the advanced age of most of the participants, this has been the only and the best way to compensate them.

Green et al. (1999) also observed in their evaluation of the project that the funds made available through the FPF appeared to have boosted the prestige of healers within the community, and the community leaders gained a better understanding of the biodiversity conservation goals and approved them. Maroons themselves seemed to understand and accept the broader ultimate objectives of the ICBG, and had interest in local economic development. This mindset of the local people was worth appreciation, and, moreover, it was also reported that there were requests from other tribal leaders to bring the project to other parts of Suriname. However, that had not been possible because of insufficient funds to expand the project.

The project evidently appears to have developed a workable mechanism and model for equitable benefit sharing, involving a single ethnolinguistic group and a low level of funding. One of the advantages of developing and testing a benefit sharing mechanism, even on a very limited scale, is that it can increase the possibility that larger sums of money will be put to good use, should they become available in future through drug royalties. The establishment and operation of the FPF provides evidence of popular participation and sustainable small-scale economic development, as well as cultural and financial appropriateness.

The Peru Project and the Maya ICBG Project (Mexico)

At this juncture, it would be pertinent to briefly discuss another ICBG project—the Peru ICBG Project—that involved Peruvian medicinal plant sources leading to the clue to great drug discovery prospects even after the end of the project period (1994–2000). Dr Walter H. Lewis of Washington University, St Louis, USA, was the group leader of this project for research on plants of the Andean tropical rainforests used medicinally by the Aguaruna people (see Box 6.1) for generations. Collaborating with local groups of Aguaruna people under the leadership of the Confederacion de Nacionalidades Amazonicas del Peru, the Universidad San Marcos, the Universidad Peruana Cayetano Heredia and Monsanto-Searle Co., a multidisciplinary team of scientists collected around 4,000 plant species. Some scientists of the original team continued the research beyond the project period by focusing on potential sources of new therapeutically active agents (Aponte et al., 2009).

Under this project, in a precedent-setting fashion, the agreements were negotiated with more direct discussions between local organizations and commercial partners (Rosenthal, 1997b) and the international scientific network that forged as a result of the project that continued beyond the project period. A large number of the species chosen, pre-screened by the Aguaruna people themselves, provided higher frequencies of bioactive secondary metabolites than those

Box 6.1:
The Aguaruna People of Peru and the Maya People of Mexico

The Aguaruna People of Peru

The Aguaruna people, also referred to as Awajun, are indigenous to the tropical rainforest of Peru. The Aguaruna communities are located historically in the geographical area of the Marañón River (a tributary of the Amazon) and its tributaries in the Peruvian Amazon. As per the Peruvian census of 2007, the Aguarunas have a population of 55,328 divided in 281 communities and located in various provinces of four different regions in Peru. They have their own language (Aguaruna) and a traditional culture, and they follow ethnic religions. They are physically well built and taller compared to the other inhabitants of Peruvian rainforest, and are known as the most skillful and brave warriors in South America. Traditionally great hunters and fishermen, the Aguaruna people are subsistence farmers of cassava, banana, peanuts and maize.

Sources: IWGIA Report: A Chronicle of Deception (2010). International Work Group for Indigenous Affairs, Copenhagen, Denmark. Available at: www.books.google.co.in/books; http://lanfiles-vm.williams.edu/mbrown/Brown-VB_GardenMagic_1980.pdf; www.mcgill.ca/cine/resources/data/awajun (All accessed in May 2015).

The Maya People of Mexico

The Maya people are the Native Americans of southern Mexico, Guatemala, Belize, Honduras and El Salvador. They historically embrace many distinctive ethnic groups. Each ethnic group has a different cultural tradition and identity and speaks a different language belonging to the Mayan language family. Despite their modernization with times, the indigenous Maya communities still preserve their identity and way of life. The community is governed by an elected 'Alcalde' that is based on a majority vote. The 'Alcalde' has complete authority to govern the villages while a relatively new village council system is in place in some villages, limiting the traditional powers of the 'Alcalde'.

Sources: www.indians.org/articles/maya-indians.html; http://digitalcommons.law.yale.edu/yhrdlj/vol1/iss1/2 (Both accessed in May 2015).

found in the flora as a whole, and the ethnobotanical focus led to pioneering legal agreements with the Aguaruna community. This experience of going beyond the project also brings to the fore that research works on nature-based products, particularly in the realm of therapeutics, still have much greater scope in future.

While the above outline of the Peru ICBG project shows the way for the future, another project—the Maya ICBG project in Mexico—initiated in 1998 under the ICBG umbrella programme was terminated in the second year of its five years of funding. According to the US Department of Health and Human Services, the project comprised a multidisciplinary programme to discover, isolate and pre-clinically evaluate pharmacologically important species from Mexico, the third richest mega-diversity region of the world and one of the most threatened in terms of biodiversity loss due to increased environmental destruction (NIH, undated). As per the project information given in this report, the species were to be screened for their activity against cancer, opportunistic diseases associated with HIV-AIDS, CNS disorders, contraception, cardiovascular disease and locally serious gastrointestinal, respiratory/pulmonary and skin disorders. The Maya-speaking people (see Box 6.1) inhabiting about 8,000 villages in the Chiapas highland are stressed by poverty, deteriorating natural resource base and extraordinary population growth rife with religious conflicts (Rosenthal, 2006).

Although the community was replete with various conflicts, the people possessed a complex ethnopharmacopoeia comprising hundreds of species of plants in an ancient system of TM. The project was a partnership of the University of Georgia, El Colegio de la Frontera Sur (ECOSUR) and Molecular Nature Ltd, a small natural products pharmaceutical and botanical company in Britain. The work progressed well initially with the aim to document the biodiversity and the ethnobotanical knowledge of the region in order to ascertain whether there were possibilities of developing medical products based on any of the plants used by the indigenous groups (Yadav and Mishra, 2013). However, the project invited severe criticisms about the methods used to attain PIC, and the probable reason could be that there were no indigenous political organizations in Chiapas highlands that existed continuously and were authorized to speak on behalf of the communities in relation to local or national resource issues (Rosenthal, 2006). It also invited the indictment of being engaged in unethical forms of bioprospecting by several NGOs and indigenous organizations. This was the first case of the kind that drew attention to the problem of distinguishing between benign forms of bioprospecting and unethical biopiracy, besides having the difficulties of securing community participation and PIC for the future bioprospectors. At last the project's partners pulled out, and the ICBG was closed down two years into its five years of allotted funding.

To summarize the experiences of the Suriname ICBG project and the brief discussion about the Peru ICBG project and the Maya (Mexico) ICBG project, we may conclude that the task of bioprospecting partnership is difficult and needs a careful tread. The 'Millennium Ecosystem Assessment' (2005) mentions that the experiences from the first 10 years of the ICBG Programme suggest that the industrial engines of innovation in drug discovery today are often small biotech companies, and many large companies are no longer screening natural product samples because they take much longer to characterize and develop than synthetic molecules from their own libraries. As a result, we find a mix of large and small enterprises in the process of development of natural products today, and this is an important lesson we learn from the ICBG programme in the context of bioprospecting. Another important lesson learnt is that if a community is not represented by some unified or single political entity, it may lead to conflicts and deter the efforts.

Case Study 3: Fund for Integrated Rural Development and Traditional Medicine (FIRD-TM) (Nigeria)

The trust fund is mostly observed as the model that appears to be more extensively accepted in different parts of the world as a tool for sharing of commercial benefits from the bioprospecting partnerships (Walter, 2002). Its mechanism has flexibility in the sense that there can be a small-scale fund, may be confined to a community level, for example, the FPF under the Suriname ICBG project, or there can be a much larger fund such as the one being discussed in this section, the FIRD-TM, Nigeria. The FIRD-TM came into being in the

Federal Republic of Nigeria, a West African country, in the year 1997, when it took shape with the announcement of a donation of $40,000 from the Healing Forest Conservancy (HFC), an independent non-profit organization founded by Shaman Pharmaceuticals, the San Francisco-based company working in tandem with Nigerian institutions, scientists and traditional health practitioners. The establishment of FIRD-TM was facilitated by the BDCP as sponsoring entity.

The account of FIRD-TM Nigeria cannot be chronicled without a brief note about of the institutions it is closely associated with. To begin with, the BDCP, a nonprofit NGO, may be mentioned here. This multiethnic international NGO was founded by scientist Dr Maurice M. Iwu in 1992. Dr Maurice Iwu was Professor of Pharmacognosy at the University of Nigeria and was the UN's Lead Consultant for the development of Nigeria's National Biodiversity Strategy and Action Plan. BDCP promotes the sustainable utilization of natural products and promotes high-quality scientific research on sustainable utilization of biological resources for health, economic development and conservation of the environment; create knowledge in the application of technology and global commerce for poverty alleviation; and develop partnerships with agencies, communities and governments for sustainable development. Another unique aspect of BDCP is that it works with women's cooperative groups, traditional healers, community groups, scientists, other NGOs, private sector interest groups, policy makers and international agencies.[5]

Interestingly, BDCP has also developed a model of ABS that is currently being used as a model for Nigeria's IPR regulations being read in the National Assembly for passage into law.[6] An innovator in ethnobotanical and bioprospecting research, the BDCP has provided consultancy services for various organizations including the African Union and the Ministry of Environment and the National Agency for Food and Drug Administration in Nigeria.[7]

The HFC, a nonprofit NGO that announced the initial donation of $40,000 to the FIRD-TM on the day of its foundation, was

---

[5] *Source*: BDCP official website: http://bioresources.org/about-bdcp/our-approach/ (Accessed in May 2015).
[6] *Source*: BDCP official website: http://bioresources.org/about-bdcp/our-approach/ (Accessed in May 2015).
[7] *Source*: www.wipo.int/ipadvantage/en/details.jsp?id=3229 (Accessed in May 2015).

created in 1990 to promote the long-term survival of the biological diversity of tropical forests, particularly medicinal plants, and the cultural diversity of tropical forest peoples, particularly their TK of medicinal plant usage. The history is that BDCP has forged imaginative joint ventures between traditional health practitioners and Nigerian scientists, and is able to facilitate a benefit sharing agreement also between the traditional health practitioners, Nigerian scientists and Shaman Pharmaceuticals. In fact, BDCP's collaboration with Shaman Pharmaceuticals was a fruitful merging of international pharmaceutical interests and goals with local scientific aspirations.

Interestingly, the initial donation from HFC was immediately complemented with pledge for additional sum by the Association of Indigenous Pharmaceutical Manufacturers and the Orange Drug Company of Nigeria. These donations indicated that the TK base and the traditional health practitioners were under the growing attention of bioprospecting institutions. In fact, the traditional health practitioners expected royalties from any commercialized product based on research inspired by their TK, and upfront payments were already being paid to them as well as to the local guides, plant specimen collectors and other helpers initially. Such payments were facilitated by the creation of the FIRD-TM. The way of functioning of the FIRD-TM is illustrated in the following paragraphs.

The FIRD-TM involves leaders of traditional healers' associations, senior government officials, representatives of village councils and technical experts from scientific institutions in managing its affairs, engaging them formally in its independent board. Its objectives largely match the objectives of the BDCP and include the programme for skill improvement to generate pharmaceutical leads that target therapeutic categories for tropical diseases suffered in Nigeria. The genesis of creation of FIRD-TM mentioned above is supported by Gupta (2002a), who observes that there has been a long tradition of bio-trade as well as bioprospecting for research and commercial purposes in Nigeria. Two major efforts have been by the National Institute for Pharmaceutical Research and Development and the BDCP, independently as well as under the auspices of the ICBG. In addition, private sector companies such as Shaman Pharmaceuticals have also been active in the country. The FIRD-TM was established as an independent body with constituents from across all sectors, including the government. This agency has a goal to build technical skills so that biological resources are a viable vehicle for improved health care and sustainable development. The fund is completely independent but may administer funds only for the purposes outlined in its charter, that is, conservation, drug development and the socio-economic well-being of rural communities.

Regarding the drug development research underway in Nigeria, it may be mentioned here that the BDCP itself is collaborating with staff from Shaman Pharmaceuticals in developing new drugs. This agency has a goal to build technical skills, using the biological resources as a means for improved health care and sustainable development. It uses local biological resources and knowledge to target therapeutic categories for tropical diseases suffered in Nigeria such as malaria, leishmaniasis and trypanosomiasis. FIRD-TM is the vehicle to receive and channel benefits in an equitable and consistent manner from many contributors, including the ICBG, of which the BDCP is a member. Funds are directed to source communities from which commercially useful biological resources and ethnobotanical knowledge are derived (Moran, 2000). The Fund was registered under the relevant laws so as to give it a legal entity capable of owning property and maintaining and defending its actions. This also helps exemption from taxes as a nonprofit organization.

The Constitution of the Fund

The initial donation from HFC was announced on the day of inauguration of FIRD-TM in the name of a pilot project to test the efficiency and efficacy of the trust fund process. The pilot HFC project generated a 21-page constitution for the fund. The HFC constitution supplies a legal mechanism to disburse widely financial resources, over a long time frame, and within varied sectors of society. The FIRD-TM is governed by the constitution that stipulates its aims and objectives, its structure, the nature of its principal organs, its financial matters and conditions for dissolution. The fund has three principal organs (Nnadozie et al., 2002):

1. **The Board of Trustees:** The property of the fund legally resides within the board of trustees; but it has no executive capacity with respect to the day-to-day running of the body.

2. **The Advisory Body:** Its capacity is purely advisory and it consists of distinguished experts in fields that are related to the objectives of the fund, as well as eminent leaders and individuals who can contribute positively to the fulfilment of those objectives.

3. **The Board of Management:** This is the executive and administrative organ of the fund. The members serve for a period of five years each and may be reappointed for a further five-year term. The Board of Management has 10 members drawn from a wide spectrum of interests and constituencies, including traditional leaders, traditional healers, government representatives and independent experts. All members are currently serving on a voluntary basis and receive no remuneration or allowance for their input except for costs incurred directly in the performance of their duties with respect to the fund.

The constitution provides for a full-time Administrative Secretary to administer the secretariat and the day-to-day business of the fund, including maintaining a record of the fund's activities and overseeing all other staff of the fund.

The Board of Trustees has among its first trustees Dr Maurice Iwu, Executive Director of the BDCP, who pioneered benefit sharing arrangements and pursued the subject of TM with great rigour, and Chief (Dr) A.A. Omotosho. The Board of Management, comprising 10 members, is responsible for managing all the affairs of the fund, including policy issues and budget, review of proposals, monitoring of projects and so on. It is chaired by His Royal Highness Eze E. Njemanze of Owerri, a highly respected traditional ruler. The BDCP can only make recommendations and it does not impose any decisions on the management committee.

The Benefit Sharing Arrangements

The Constitution provides for the distribution of benefits to all stakeholders in accordance with the following guidelines (Moran, 2000):

1. At least 50 per cent, but not more than 70 per cent, of available funds shall be distributed to traditional healers' organizations and community development funds.

2. At least 10 per cent, but not more than 15 per cent, of available funds shall be distributed to national universities and other national institutions that share a commitment to the aims and objectives of the Fund.
3. At least 10 per cent, but not more than 15 per cent, of available funds shall be distributed to the sponsoring entity for its furtherance of conservation and development activities.

In the WIPO–UNEP study about the functioning of the fund, Professor Anil K. Gupta (2002a) found the following general principles for allocating financial resources from the fund:

- About 60 per cent was kept in fixed deposit of which only the interest to be used.
- About 40 per cent of the fund, that is, $40,000, was set aside to be used during 1999–2000. Of the 40 per cent:

    i. 20 per cent was to be used for the Biodiversity Conservation Act of the National Institute,
    ii. 10 per cent for educational purposes,
    iii. 30 per cent for the Traditional Healers' Association for group projects or micro-credit funding,
    iv. 30 per cent for community development associations for village projects,
    v. 5 per cent for women, especially widows, and
    vi. 5 per cent for children's welfare.

At the user end, regarding projects that can be taken up under the fund, town associations, village heads and professional guilds of healers are empowered to make decisions in their localities. The fund is aimed at enhancing the capacity of traditional health practitioners and is expected to support local projects, help build herbal clinics and botanical gardens and monitor various activities. There is a view that it should not focus on helping only individuals but also help communities (Gupta, 2002a). In his study, Professor Gupta enumerates the various stakeholders in the benefit sharing chain in FIRD-TM as the individual healers who provide knowledge, the communities that may provide leads or conserve the biodiversity, the association of healers who help in maintaining professional quality and responsibility, the scientists in Nigeria and the scientists in the USA (in BDCP as well as in Shaman Pharmaceuticals or Walter Reed Army Medical Research Center).

The experiment of the FIRD-TM in Nigeria has a history. As Moran (1999) assuredly claims that in practice, a case study of how countries, companies and cultures can cooperate is that of Shaman Pharmaceuticals, Inc. in Nigeria. She narrates that Dr Maurice Iwu and another Nigerian scientist, Cosmos Obialor, proposed initial discussions with healers and traditional leaders to talk about a collaborative relationship with Shaman well before 1992, the year the CBD was introduced. They visited communities, and after lengthy discussions, the groups felt that Shaman shared a common purpose with them consistent with their cultural values concerning human health. Out of these discussions, the PIC and compensation policies of the company were formulated. PIC discussions covered topics such as the intentions and goals of the project, how and where the plants would be analysed, their potential for commercialization and benefit sharing. The company reports laboratory results back to the participating communities regularly. General literature is published on medicinal plants from Nigeria, supplying public recognition of the benefits of TK from Nigeria to society and human health.

While discussing the fund for managing the TM, it would be apt to look at this aspect also from historical point of view. Professor Gupta (2002a) in the WIPO–UNEP study describes the history of TM in Nigeria and mentions that there are two primary communities with whom BDCP works closely in this project. The first one is the community of traditional healers through their national union as well as state-level associations. The second one includes the communities at the village level where conservation, biodiversity monitoring, sustainable extraction and other activities related to rural development are being planned. Professor Gupta (2002a) states:

> Nowadays, the traditional healers take notes and write up their diagnosis, whereas earlier they did not do that. The knowledge was treated as a common property and shared widely, though there are healers who keep it secret.

And yet there is no centre for medical research in Nigeria managed and owned by traditional health practitioners except the one set up by BDCP. The clinic set up in 1992 as a result of the BDCP's involvement with traditional healers is owned and managed by the local traditional healers themselves.

Commentary on the Fund

It is interesting to learn that the replication world over in establishing trust funds as the model of benefit sharing is increasing and is being studied also by many agencies. Gupta (2002a) points out that there have been some tensions on the issue of the allocation of funds to the communities vis-à-vis the healers' union. The decisions so far have favoured the healers' unions partly because they are better organized and are also well represented in the Board. However, healers' organizations, informal as they are, may generally not be able to fulfil all the requirements for a community project—that it should have some kind of organization, bank account, list of the key members, nature of activities and duration, apart from the local contribution towards employment and development. The project format also requires information about the guarantor.

The Fund offers an example of how countries, culture groups and companies can work together successfully for the benefit of all stakeholders to sustainably develop biodiversity for human health. Moran (1999) adds that the trust fund concept offers the added value of attracting and managing sources of financing from other NGOs, foundations or companies interested in contributing to a stable fund. Her view is supported by the additional donations that came from the Association of Indigenous Pharmaceutical Manufacturers and the Orange Drug Company of Nigeria to complement the initial donation from HFC. Another strong point of the fund is highlighted by Gupta (2002a) when he mentions that the sharing of benefits by Shaman Pharmaceuticals through BDCP even before any drug was developed or even in the case where they have used public domain TK is considered to be a good practice. The commitment to share benefits with the communities from whom Shaman has sourced any material at any time, even if the final products emerged from only one lead provided by only one community, is also a novel idea. However, the functioning of the FIRD-TM is not all rosy in nature as many commentators state that several questions remain unanswered, such as striking an appropriate balance between the interest of organized groups like traditional healers and that of unorganized groups like local communities and unregistered healers. Professor Gupta (2002a) reiterates that the voice of traditional healers has received greater recognition than the voice of local communities and their leaders, and the local communities are generous in sharing their knowledge and yet hardly any publications were found in which inventors were given joint authorship.

The reservations mentioned above might have been taken care of by the fund managers by now, and conclusively, a statement about the FIRD-TM would be appropriate to be quoted here (Nnadozie et al., 2002):

> It (FIRD-TM) was a response to the major institutional gap and the lack of an appropriate and effective vehicle to receive and channel benefits in an equitable and consistent manner in order to source communities from which commercially useful genetic resources and specialised knowledge are derived.

Besides the advantage that trust funds can range in size and scope, there is flexibility of administering these funds at different levels as per the specific requirements. According to Guerin-McManus et al. (2002), these funds provide a stable and enduring structure that can last over long periods of time for the purpose of channelling benefits in a controlled and consistent manner. The advantage with trust funds is that they aim at avoiding problems associated with direct cash payments to individuals or communities. No matter what, the experience world over in establishing trust funds as the models of benefit sharing is increasing and still they are being studied by many agencies.

Case Study 4: Benefit Sharing with the Kani People (India)

This example relates to the benefit sharing arrangements arrived at between the Kani people of Kerala (India) and the Tropical Botanic Garden and Research Institute (JNTBGRI)[8] for the development of

---

[8] TBGRI is renamed as JNTBGRI. It is an autonomous institute under the Kerala State Council of Science, Technology and Environment, Government of Kerala (India). Website: www.jntbgri.in (Accessed in May 2015).

a TK-based drug called *Jeevani* that was developed from a local herb and was licensed to a pharmaceutical company with a scheme of benefit sharing involving the Kani people.

The JNTBGRI developed this energy-giving drug, making use of the TK provided by the Kani people, and obtained a patent for the drug. Licence was granted for the manufacture and sale of this drug for a fee of ₹10 lakh to Arya Vaidya Pharmacy Ltd. (AVP), a Coimbatore-based pharmaceutical company pursuing the commercialization of Ayurvedic and herbal formulations since 1948. The JNTBGRI agreed to share 50 per cent of the fee and royalty from the drug with the Kani people through a trust fund for development and biodiversity activities (Swiderska, undated). The drug was developed from the leaves of a plant commonly known as Arogyapacha (*Trichopus zeylanicus*), a small rhizomatous, perennial herb distributed in Southern India as well as in Sri Lanka and Malaysia. Within India, the plant is endemic to the region of the Western Ghats that falls in the Thiruvananthapuram district of Kerala and the Tirunelveli district of Tamil Nadu. The development of this anti-fatigue drug marks a typical case of benefit sharing. This case manifests a composite of success and inherent conflicts, and in that respect, it makes a brilliant example of benefit sharing, which has been studied widely and keenly. The story is highlighted in the following paragraphs.

The Kani people ate the fruits of the Arogyapacha plant for its anti-stress and anti-fatigue properties during long treks in the hills. This was revealed in 1987 during field surveys conducted as part of a research work, under the All India Coordinated (Research) Ethnobotany Project (AICEP), to Dr P. Pushpangadan, who had come down for the field survey from the Regional Research Laboratory, Jammu, where he worked before taking over as the Director of JNTBGRI in 1990. Taking a lead from this, Pushpangadan and his team of scientists isolated 12 active compounds from Arogyapacha, developed the drug *Jeevani* and filed two patent applications on the drug (Gupta, 2002b; WIPO, 2003b).

The forest where the Arogyapacha plants are found is the Augustayar or Agasthyamalai forest of Western Ghats, a reserved tropical rainforest. It so happened that during an ethnobotanical expedition to the Western Ghats, the Kani tribals working as guides on the hilly trek looked energetic and agile, whereas Pushpangadan and other scientists of his team felt exhausted. The Kani guides were continuously munching black fruits of some plant, which they offered to the scientists too, and the team felt full of energy and vitality. It was then that the Kani guides, after great reluctance, revealed the identity of the Arogyapacha plant, and thus unfolded this story of ABS subsequent to the shifting of research work on the plant coinciding with moving of Dr Pushpangadan from Jammu to JNTBGRI.

The Drug Development

The chemical screening tests and other investigations for isolation of the active ingredients of the plant were taken up by Pushpangadan along with Avinash K. Sharma and C.L. Chopra at Regional Research Laboratory, Jammu. The scientists carried out chemical and pharmacological investigations of the Arogyapacha plants, and it is likely that only *Trichopus zeylanicus travancoricus*, the species found in Augustayar (also known as Agasthyar), has the claimed medicinal properties, though the plant is found in Malaysia and Sri Lanka as well (Gupta, 2002b). Initial investigations revealed the presence of certain glycolipids and non-steroidal compounds that possessed anti-stress, anti-hepatotoxic, immuno-restorative and anti-fatigue properties, revealing much greater potential than what the plant was being used for till then. During the last phase of the research at the JNTBGRI, the drug Jeevani was formulated, using this and three other medicinal plants[9] as the ingredients (GOI, 2002).

The analysis included both allopathic and Ayurvedic tests, and later the relevant clinical trials also followed, which were highly significant. Two of the patent applications on *Jeevani*, as stated in the WIPO–UNEP study by Gupta (2002b), were for:

1. Process of preparation of novel immuno-enhancing anti-fatigue, anti-stress and hepato-protective herbal drug (Pushpangadan, P., Rajasekhran, S. and George, V., 1996, Patent application number 959/MAS/96 dated 4 June 1996);

---

[9] *Withania somnifera* (Ashwagandha), *Piper longum* (Pippali) and *Evolvulus alsinoides* (Dwarf morning glory, also is one of the four herbs that are given the common name of Shankhapushpi).

2. Process for the preparation of a glycolipid fraction from *Trichopus zeylanicus* possessing adaptogenic activity (Butani, K.K., Gupta, D.K., Taggi B.S., Anand K.K., Kapil R.S., Pushpangadan P. and Rajsekhran S., 1994, Patent application number 88/Del/94); and

3. Two patent applications that included *Trichopus zeylanicus*, one for diabetes (957/MAS/96, dated 4 June 1996) and other for a sports medicine (958/MAS/96 dated 4 June 1996).

The drug has already drawn attention the world over and seems to have potential in the international market as a natural health care and sports medicine. And yet, the research is not complete. The JNTBGRI isolated five compounds in all from Arogyapacha, but the lack of technology forced the scientists to send four compounds to Copenhagen for detailed characterization. In order to bargain from the wonder drug, following the guidelines of the CSIR, the JNTBGRI licensed the technology for manufacturing *Jeevani* to AVP, transferring the right to manufacture the drug vide an agreement signed in 1996 for a period of seven years. In addition to the license fee of ₹10 lakh, the JNTBGRI was also to receive royalty of 2 per cent on any future sales of the drug. The JNTBGRI emphasized that the licence period was only for the purpose of a promotional venture, and that once the drug was able to establish a market for itself within the licence period of seven years, the licence fee could be suitably enhanced and that it could be licensed to another company if that was more beneficial (Gupta, 2002b).

These benefits were accruing from the TK held by the Kani people, and therefore the JNTBGRI decided to share the benefits with the Kani community in general, with substantive share going to the Kani guides who provided the clue that led to the development of the drug. As part of the arrangement for benefit sharing, the JNTBGRI decided to share half of the licence fee as well as the future royalties from the sale of the drug with the Kani people.

It was this idea of benefit sharing that culminated into the setting up of a trust fund to institutionalize the benefit sharing process and also to administer programmes for the socio-economic development of Kani community and also for related biodiversity conservation activities. On the initiative of the JNTBGRI, the trust named as KKSK Trust was set up in 1997 with nine members from the Kanis. Within its structure, the trust includes representatives of 30 Kani settlements and the president and vice-president of the trust are the two Kani guides who gave clue of *Jeevani* drug to the JNTBGRI scientists.

The deed of the trust states the following objectives:

1. To administer activities for community welfare and socio-economic development for the Kani people;
2. To undertake the preparation and maintenance of a biodiversity register to document the Kani's knowledge base of plants and animals; and
3. To evolve and support methods to help the local communities promote sustainable use and conservation of biological resources.

The capital received by the KKSK Trust is maintained, with interest used for community development projects. The objectives include, as part of the welfare and development projects, activities such as setting up of a telephone booth in a remote area, insurance schemes for pregnant women and another covering accidental death. Also included are support mechanisms for poorer members of the community. The benefit sharing model covered over 700 families.

The Conflicts

There are several aspects in this study that need a closer look. Awareness among the Kani people about the KKSK Trust was very low, and there was some criticism of the mode of sharing of the benefits. Because of the protests of some Kani groups, the disbursement of money could not take place for some years. Members of the trust were of the view that once the trust became functional, they would be able to organize the Kanis better.

The work of the KKSK Trust and its cooperation with the JNTBGRI led to employment and income generation activities related to the cultivation and post-harvest processing of Arogyapacha. Training was provided by the JNTBGRI to ensure that, while an adequate supply of raw material was harvested, techniques were sustainable and the wild plants were not over-harvested. The initiative empowered the tribal community to protect, preserve and maintain their knowledge, innovations and practices of conservation and sustainable use of biological resources.

The only drug *Jeevani* that has been developed seems to have potential in global markets as a natural health care and sports medicine. The AVP was disappointed that despite there being a good market for the drug, only within few years of its launch, there was no adequate raw material to continue the manufacturing of the drug. However, the propagation of this plant in moist and shady environment is quite easy, and many Kani tribals have actually already cultivated this plant. A pilot project for cultivation of the plant was carried out during 1994–96 with support from the Integrated Tribal Development Programme in areas surrounding the reserved forests. Scientific studies are noted to reveal that the medicinal properties of the plant are best manifested in plants growing in the natural habitat. Under the arrangement, 50 families were given around ₹2,000 each for cultivating the plant, and the JNTBGRI was to buy the leaves to supply them to AVP. But further cultivation of the plant could not take place as the plant was not in the Forest Department's notified list of minor forest produce (GOI, 2002).

The Kerala Forest Department was concerned about the sustainable extraction and thus had not allowed this plant to be commercially exploited so far. It feared that commercial interests could pressurize the Kanis towards indiscriminate exploitation of the plant from the forests, as cases of smuggling of the plants had already begun to surface (GOI, 2002). A recent report indicates that the Forest Department has now consented to consider the plant in its list of minor forest produce and make arrangements so that AVP can buy the leaves directly from the Kani people (GOI, 2002).

Some questions that arise from this case study are: What kind of legal instruments are available for protection of the TK associated in this arrangement? What role should IPRs have in the arrangements of benefit sharing associated with TK? How should benefit sharing be structured between the company that developed the product and the tribe that conserved and that cultivates the plant and holds the knowledge about it? Would the access to the genetic resource have remained free from regulation, if AVP had been located in another jurisdiction? How should such access be regulated, and what should be the role of the community and the state? Several other aspects such as ownership of the patent, the proportion of benefits that should go to different clans of Kanis, sustainable management of Arogyapacha plants as a resource and domestic cultivation and sustained supply are still under close investigation, and the nitty-gritty of actual sharing of benefits have yet to be given practical shape.

Anuradha et al. (2001) in their case study supported by the UK Department for International Development (DFID) have highlighted the conflicts associated with the endeavour of the JNTBGRI with the Kani people of Kerala. They state that the benefit sharing arrangement has been through some 'chaotic phases' over the past four years. The issues involved in this conflict as described by these authors are as follows:

- The Forest Department has not permitted the Kanis to collect Arogyapacha, and hence there has been a 'stalemate on the issue of collection of the leaves' and thereby manufacture of the product for the past five years.
- There is no uniformity in the Kanis' perception of the benefit sharing arrangement proposed by the JNTBGRI, as the Kanis are no longer a single cohesive unit or community.
- The JNTBGRI believes that the formation of the KKSK Trust would result in greater involvement of the Kanis in the benefit sharing arrangement. However, Kanis from a number of areas are yet to become part of the Trust.
- A concern is raised about the lack of coordination and mechanism for dialogue between the JNTBGRI, the Forest Department and KIRTADS (Kerala Institute for Research, Training and Development Studies of Scheduled Castes and Scheduled Tribes, an official agency—a research institute for tribal studies), which are all part of the same government.
- A host of complexities need to be resolved, including who owns the resources and related knowledge when such resources occur on private land, public land or both types of land; and when the resources and related knowledge occur in more than one state.

The discord with the Forest Department has been narrated by Gupta (2002b) in his study for WIPO–UNEP, showing the Forest Department's concern about sustainable extraction of Arogyapacha and thus not allowing its commercial exploitation so far. Based on his discussions with the Forest Department in 1999, Gupta suggests that the JNTBGRI should agree that any drug that it develops from forest-based plants should be licensed for commercial use only under three conditions:

1. All the four parties, that is, the JNTBGRI, the Forest Department, the local community institutions and the licensees, should be involved in the discussion.
2. A sustainable extraction plan should be submitted by the licensee to ensure that commercial utilization does not pose any threat to the ecosystem or long-term sustainability of the species.
3. Research programmes on such plants should be reviewed by the JNTBGRI and the Forest Department from time to time so that if any endemic, endangered plant provides a lead for a valuable medicine, then unless technologies are developed for ex situ cultivation through tissue culture, such a technology will not be commercialized and licensed lest the plant becomes extinct.

It may be highlighted in this case that the biological resource from which the benefits were being drawn was being harvested from forest land where the Kanis had no tenurial rights. The forest land belongs to the Kerala Forest Department, and there was no objection to the harvesting of Arogyapacha leaves initially. The Kerala Forest Department stopped the people from harvesting the leaves on the ground that the exploitation had become commercial and was threatening to the very existence of the species. Ironically, the Forest Department (though very much an on-site stakeholder), not being a party to any of the agreements, did not try to evolve any mechanism for sustainable harvesting of the resource, adopting the easy recourse of putting a ban. Later, some efforts were made to raise Arogyapacha plants through cultivation, but the efforts were unsuccessful.

Another issue of conflict was that the JNTBGRI was negotiating with the Kanis from a particular belt, and the Kanis from other adjacent areas were up against the JNTBGRI on the ground that the JNTBGRI would have consulted them or their medicinal practitioners, who are called as *plathis* locally, before making any benefit sharing arrangements.

Commentary on the Attempts Made

The case of *Jeevani* underscores the complexities of attempting to establish a voluntary ABS agreement and highlights some of the pitfalls associated with such schemes (WIPO, 2004b). Commenting on the 'limited success' of the efforts, Utkarsh et al. (2005) state that the JNTBGRI acknowledged in writing, if not through a formal contract, the initial contribution of the Kani tribals and offered to share with them half of the royalty. They observe that in the absence of appropriate policy guidelines and government channels, the money has not yet reached the tribals even a few years after the JNTBGRI handed it over to the government for the purpose of the tribals. The production of the drug has also run into problems as the raw material (the plant) is declared as rare and its exploitation is banned by the Forest Department. The cultivation efforts are minimal, and their success, especially chemical yield, is limited. Therefore, the well-intentioned and much appreciated effort, and probably the single endeavour of its kind for benefit sharing in India that was initiated before appropriate policy and legal framework, had only a limited success.

WIPO (2004b) highlights the following aspects in this case:

- The benefit sharing arrangements in this case highlighted the potential value of an active role by local authorities in partnership with community leaders.
- While the establishment of the Trust Fund has been widely applauded, it could have ensured wider involvement from all sectors of the Kani communities.
- One of the biggest problems has been the shortage of leaves from which to process the *Jeevani* products. Tests indicate that varieties found wild in forests yield the best compounds.
- If the Forest Department had been involved from the beginning, perhaps their attitude might have been different.

In another study of the endeavour of the scientists of the JNTBGRI involving process patent of *Jeevani* and the Kani's trust fund, Gupta (2003) has analytically commented as shown in the following points:

- If the Forest Department has the legal control of a territory, then one cannot ignore them while establishing benefit sharing mechanism.
- The fact that scientists did not claim any share from the licence fee only goes to prove what kind of values and motivations guided this benefit sharing model.
- The degree of involvement of various tribal settlements and groups could have been increased and the rights of informants vis-à-vis the communities needed more discussion among the communities themselves.
- The non-material contribution of benefits by way of empowerment of local communities deserved to be noted, but several more such benefits could have been considered. For instance, the health check-up of local communities was urgently needed, given a very poor condition of many women, children and also some male adults among Kanis.

The story of Kanis does not end here with *Jeevani*. The Kerala Forest Department and the JNTBGRI are preparing to file patent applications for three herbal products that were developed, based on the TK of the Kanis, during a World Bank-aided pilot participatory project on the conservation and sustainable use of MAPs (*The Hindu*, 2004). This project also establishes a biodiversity register and a database on indigenous knowledge systems.

In a nutshell, the issues that have yet to be resolved are: How and in what proportion the benefits should be distributed among the different clans? How can the resource management be rendered sustainable? Is it feasible to domesticate the species, and cultivate it intensively? What are the steps involved and how an approach may be made to give a practical shape to the benefit sharing model in order to satisfy all the stakeholders—the *plathis*, researchers and the community—equally well? Now it is gathered that the Kerala Forest Department is planning to invigorate the endeavour and come up with a sustainable project that would make another beginning of the production of *Jeevani* by involving all the major stakeholders including the industry and JNTBGRI. The composite of success and inherent conflicts of this case, as stated in the beginning, make it unique also in the sense that the arrangements of benefit sharing in this example can be a guiding and motivating factor for many more of such endeavours, for the arrangements were in place much before a formal ABS regulation came into force in India in the form of the Biological Diversity Act of 2002 and the Biological Diversity Rules of 2004.

Case Study 5: San Hoodia Case of Benefit Sharing (Southern Africa)

A well-studied case of benefit sharing, the 'San Hoodia case' from southern Africa, is illustrated and evaluated in this section. The case is a wonderful example in the context of benefit sharing wherein the indigenous community that held the TK from which the benefits accrued had to employ a lawyer and had to put a fight to an unwilling government-owned institution in order to share the benefits. The case is illustrious even in the sense that the indigenous community was not well defined, not educated, lived in abject poverty and crossed national boundaries spanning four countries of southern Africa.

The product in focus is an appetite suppressant extracted from Hoodia (*Hoodia gordonii*), a leafless succulent plant native to the Kalahari Desert in southern Africa. The plant's vernacular name is *Ghaap*, which is used commonly for *Hoodia gordonii* and also for *Trichocaulon piliferum*, now known as *Hoodia piliferum* (Wynberg, 2004), both of the species having similar properties, though the latter does not appear to draw the attention that the former draws. The fleshy stem of Hoodia was chewed by members of the San community of southern Africa on their long hunting and gathering trips in the deserts to quench thirst and to suppress hunger. The San people are thousands of years old natives, also known as 'Bushmen' or *Basarwa*, whose population is now limited to less than a hundred thousand (Suzman, 2001), spread over Botswana, Namibia and South Africa with scanty populations in Zimbabwe, Angola and Zambia.[10] It was

[10] The country-wise population figures of San people, according to an assessment made by J. Suzman (2001), with their percentage to total population of the country given in

a centuries-old TK that was not shared with the outside world before the year 1930. It so happened afterwards that the San people shared their TK with a Dutch anthropologist who published his findings in a book, and following the clue, much later in 1963, the South Arica's CSIR began carrying out R&D into the properties of Hoodia (WIPO, undated). The scientists of CSIR could finally identify the active ingredient in the plant, named it as 'P57' molecule, and got it patented in 1996. From here unfolds the engaging and now prominent story, chronologically narrated below, which became the subject of about 10 documentaries, more than a dozen PhDs and master's dissertations registered to investigate it further and hundreds of news items written (Wynberg and Chennells, 2009). In fact, the lack of technology within CSIR to isolate and identify the active ingredient effecting the appetite suppression in Hoodia was the reason for insufficient evidences that were essential to file for a patent (Wynberg and Chennells, 2009). Their project got impetus in 1986 when they acquired suitable gadgets to conduct high-field nuclear magnetic resonance spectroscopy (Wynberg, 2004; Wynberg and Chennells, 2009), and a decade later, they could obtain the patent for P57.

For the commercial uses of the newly patented molecule, CSIR signed a licence with Phytopharm, a small British company dealing in phytomedicines, in 1998. Simultaneously, CSIR was granted patents in the USA and some other countries. The things began moving fast from here, as having an exclusive worldwide licence to manufacture and market Hoodia-related products and to exploit any other part of the CSIR's IPR relating to Hoodia species, Phytopharm developed the drug lead 'P57' to a more advanced stage, leading to a licence and royalty agreement with Pfizer, the US-based pharmaceutical giant, for further development and commercialization (Wynberg and Chennells, 2009). Due to some complexities involving clinical trials, Pfizer discontinued the licence in 2003, but fortunately, in the following year, Phytopharm granted an exclusive global right to Unilever, a food multinational giant, to market Hoodia-based products as anti-obesity drugs and also as functional foods (herbal and dietary supplements) for its usefulness in other metabolic disorders. From here

parentheses, are: Botswana: 47,675 (3.30), Namibia: 32,000 (1.80), South Africa: 4,350 (<0.02), Zimbabwe: 2,500 (<0.02), Angola: 1,200 (<0.01) and Zambia: 300 (<0.01).

there was no looking back for the CSIR, and surprisingly, the San people remained quite unaware that without their consent or their sharing of the benefits derived from their TK, an institution was on its way to a large-scale commercialization of Hoodia-based phytomedicines. The further development in the direction of benefit sharing arrangements between the San people and the CSIR is taken up in the following paragraphs.

The international print media and some of the South African NGOs were watchful about the new kind of development taking place in South Africa as the debates with reference to the CBD had already brought in adequate awareness about the benefit sharing issues. The British newspaper, *The Observer*, published a story (*The Observer*, 2001), mentioning that 'Hoodia is at the centre of a bio-piracy row'. The Biowatch South Africa, an NGO, assisted by Action Aid, an international NGO, also alerted the foreign media to the potentially exploitative nature of the CSIR–Phytopharm agreement (Wynberg and Chennells, 2009). The issue was taking an interesting turn and the patenting of P57 by CSIR came into limelight. Along with these developments, the San community also was slowly gaining strength with time, uniting across the national boundaries. In the year 1996, the San people had formed the Working Group of Indigenous Minorities in Southern Africa (WIMSA) with an aim to represent the San communities from Botswana, Namibia and South Africa. San leaders in WIMSA ensured that their cultural and linguistic diversity was celebrated under a collective San cultural umbrella, which proved decisive in their aim to achieve San unity across national boundaries (Wynberg and Chennells, 2009). It seemed that the San people were about to determine their fate now. They formed the South African San Institute (SASI), an NGO with its base in Cape Town to facilitate funding and expertise to them and to partner with the WIMSA and other NGOs to campaign for them (WIPO, undated).

The Negotiations for Sharing of Benefits

As the luck would have it, the San people came in contact with an excellent lawyer, Roger Chennells, working with the human rights law firm Chennells Albertyn, who was engaged to act for WIMSA and SASI, in representing the various San peoples in South Africa in the Hoodia case (Tellez, undated). In 2001, as part of WIMSA, a voluntary association known as the South African San Council was established by the three San communities of South Africa. These new institutions and the active participation on behalf of Roger Chennells led to build up pressure on CSIR to accept the lapses and agree to share the benefits that were being accrued from the TK of the San people. Still, as the San people as well as the Hoodia habitat were not confined within South Africa only, the CSIR was reluctant to negotiate with parties outside the country, so, through WIMSA, the South African San Council was formally mandated to represent the San of Namibia and Botswana as well as those in South Africa, and WIMSA and SASI instructed their lawyer to negotiate with the CSIR on behalf of the San (Wynberg and Chennells, 2009). Ultimately, the discussions between the two parties began and, after months of negotiations, a memorandum of understanding (MOU) was reached in 2002.

The process was controversial and involved legal negotiations through meetings and workshops. At last, the CSIR had to acknowledge the role of the San people, their TK and innovative activity in the initial discovery and development of the properties of Hoodia. The significant features of the MOU signed between the CSIR and the South African San Council included that any IP arising from the traditional use of Hoodia and related to the CSIR patents for P57 remained vested exclusively with the CSIR, and the San Council had no right to claim any co-ownership of the patents or products derived from the patents. Both the parties agreed for complete transparency in future negotiations and committed themselves to be in good faith in order to arrive at a comprehensive benefit sharing agreement.

The parties reached the final agreement for benefit sharing in the following year. The key provisions in the benefit sharing agreement were:

1. The San communities to receive 8 per cent of all the milestone payments[11] that are received by CSIR from the licensee firm, Phytopharm, so long as the drug is in clinical development over the forthcoming years.

---

[11] Payments to be made during the clinical development stage, whenever certain technical performance targets are completed successfully, being marked as milestones.

2. Of all royalties received by the CSIR from Phytopharm as a result of the successful exploitation of products, 6 per cent to go to the San people for the duration of the royalty period, or as long as the CSIR received financial benefits from commercial sales of the products.
3. Any IP arising from the use of the TK related to Hoodia and from the patents for P57 to remain vested exclusively with the CSIR, and the San Council to have no right to claim any co-ownership of the patents or products derived from the patents.
4. Both the parties to conserve the biodiversity and undertake best-practice procedures for plant collection.
5. The CSIR to lay the groundwork for further collaboration in bioprospecting.

The San Hoodia Benefit Sharing Trust

An initial milestone payment of $32,000 has already been made to the San Council. In the follow-up of the agreement reached for benefit sharing, a trust was formed and registered in the year 2005 as the San Hoodia Benefit Sharing Trust (or 'San Trust'). The San delegates from South Africa, Namibia and Botswana also participated in the debates and agreed on the issues relating to the principles of benefit sharing. It was agreed that of the Trust income would be allocated as shown below (WIPO, undated):

- 75 per cent would be equally distributed to the constituted San Councils of Namibia, Botswana and South Africa.
- 10 per cent would be retained by the San Trust for internal and administration purposes.
- 10 per cent would be allocated to WIMSA as an emergency reserve fund.
- 5 per cent would be allocated to WIMSA to cover administration of the San networks.

The founding of the San Trust basically aimed at proper channelling of fund flow and a set up for prioritizing education, financing San organizations and community projects, and also for the purchase of ancestral lands that were taken away from the San. The San people also planned to invest the money, and only tap into the interest generated to fund community projects. It was also provided that no payments would be made to individuals, and that both the CSIR and the San Trust would adopt clear and transparent accounting procedures in place with regard to the use of the financial benefits by the San Trust. According to Wynberg and Chennells (2009), it was also propounded that no distribution of funds would be made to a beneficiary community or institution unless there was a request approved formally by the San Trust and detailed budget and coherent plan were set out besides the strict rules framed for the distribution of funds.

A proper constitution and elected representatives along with opening of a bank account were also provided. There was also an arrangement to share money from the San Trust with various San organizations to sustain their programmes in Angola, Namibia, South Africa and Botswana. Regarding the membership, it was agreed upon that the San Trust would include representatives of the CSIR, all the San stakeholders in southern Africa, WIMSA, a South African lawyer nominated by the South African San Council and the Department of Science and Technology, with strict rules determining the distribution of funds to beneficiaries (Wynberg and Chennells, 2009).

Commentary on the San Hoodia Case

The San Hoodia case stands out in the sense that an impoverished community, which spanned across boundaries of nations, has been successful to win back the very legitimate rights of its own TK, and a government institution (CSIR) was involved in a case akin to biopiracy while the institution should have been acting 'for' and not 'against' the sharing of benefits. It can only be said that whatever kind of 'biopiracy' occurred, it occurred before South Africa enacted its Biodiversity Act.[12] This case also reflects the scenario wherein the

---

[12] South Africa promulgated the Biodiversity Act in June 2004 that has a provision for 'fair and equitable sharing among stakeholders of benefits arising from bioprospecting involving indigenous biological resources'. It also has provisions requiring prior informed consent

international media and NGOs were awakened enough to highlight a case that was against the spirit of the CBD. Besides, the provision that some part of the fund would be used for the purchase of ancestral lands that were taken away from the San is an indication of emphasis on empowerment of the San people.

In South Africa, Hoodia has been put to cultivation practices employing local San communities, which reflects brighter prospects as the cultivation could be the main source in the commercialization of P57 molecule (WIPO, undated). In the light of the fact that the threats posed to natural growth of Hoodia through unregulated collection led to the inclusion of Hoodia species in Appendix II of Convention on International Trade in Endangered Species of Wild Fauna and Flora (CITES), the provision in the agreement for conserving biodiversity and undertaking the best-practice procedures for plant protection may prove to be useful and supportive in the long run.

Another novelty recounted by Wynberg and Chennells (2009) is that Hoodia being a biological resource shared across national boundaries, South Africa played a leading role in lodging the patent, developing commercial partnerships with multinational companies, negotiating benefit sharing arrangements with the San and facilitating legal trade in the plant while Botswana and Namibia, by comparison, although involved in harvesting and cultivating Hoodia, have not yet legalized trade in the plant nor developed commercial partnerships. Furthermore, Namibia and Botswana, unlike South Africa, have no ABS regulations, and if at all the royalties start pouring in, it will lead to new intra-community conflicts, making it difficult to determine how equitable benefits would be spread within communities and also across the national boundaries. How the subsequent negative impacts would be dealt with becomes a big question. This issue was raised by some commentators, for the TK related to Hoodia was shared among a broad spectrum of communities—the non-San groups—known as Nama, Damara and Topnaar who also have been occupying the Hoodia growing areas.

---

to be obtained from the holders of TK for use of such knowledge in bioprospecting and for benefit-sharing agreements with them.

Some of the challenges we find in this case study include the one that the San people are not receiving any revenue from the sales of many Hoodia-based products currently being traded in the international market because such products are being commercialized outside of the CSIR agreement (Tellez, undated). The reason, some experts state, is that CSIR has patented the P57 molecule only, while dozens of new Hoodia-based products are sold as functional foods.

It may be stated in a nutshell that the San Hoodia case serves as a classic example for benefit sharing agreements that were reached after a fight an indigenous community had to put to a government-owned institution, and one can be optimistic that cross-border benefit sharing arrangements also may become a possibility.

Case Study 6: The Argan Oil Commercialization (Morocco)

This case of benefit sharing is only one of its kind in the sense that it is reported as having adopted a CSR approach on behalf of the main resource user, a multinational cosmetic industry. The unique case from Morocco, a North African country that still does not have any ABS legislation, involves the export of argan oil and oil cakes along with some other products to the cosmetic industry in many developed countries, and at local level, it involves all-women cooperatives, with rural women being the main actors in the extraction of the oil.

The argan oil received the utmost recognition during the 1990s for its nourishing, cosmetic and medicinal values, making it the world's probably most expensive edible oil. The processing and marketing of argan oil has expanded at global level to a large extent during the recent past. Especially for its cosmetic uses, it sells at much higher price and has become the subject of several US and European cosmetic patents (Lybbert et al., 2011). The oil is a delicacy as mixed with almond and honey, it makes a tasty spread to be put on bread to welcome guests, and it is often used as the first feed of the newborns. It is also used as a salad dressing or added to dishes after cooking (Guillaume and Charrouf, 2011). Because of its ever-rising prices and augmenting the household income of the local inhabitants, it came to be known as the 'liquid gold', and the argan products represent a major income for about 6 per cent of the rural population and up to 90 per cent of the economy in areas of native argan stands (L'Oréal Canada, undated; Lybbert et al., 2010). The dependence on argan agroforestry is governed by clear and well-established, albeit complex, tenure arrangements, while some have argan trees on private land, most households access argan fruit via seasonal usufruct rights in defined forest tracts called *agdal*, and after fruit harvest, these usufruct tracts return to collective exploitation (Lybbert et al., 2010).

The argan tree (*Argania spinosa*) is endemic to Morocco, growing only in the south-west of Morocco in an area covering about 870,000 hectares, about 7 per cent of the forest resources of Morocco (Taleb, 2014), and is little known outside Morocco. The trees can grow up to 8–10 m in height and live from 150 to 200 years. Argan trees are resistant to drought and heat, growing naturally in the arid and semi-arid regions, and within the argan region, the trees play a vital role in the food chain and the environment. Expressing his concern about the declining argan ecosystem in Morocco, Mohammed Sghir Taleb, a research scientist with the University Mohammed V in Rabat, Morocco, while presenting a study of argan agroforestry during the World Congress on Agroforestry in New Delhi (10–14 February 2014), stated that the argan agroforestry system sustains an estimated three million people, and in less than a century, argan tree density has decreased from 100 to 30 trees per hectare due to unsustainable exploitation (Lybbert et al., 2010; Taleb, 2014). This is a matter of great concern, as Taleb states that the argan ecosystem provides an important source of timber, forage and firewood for local communities, but the ecosystem is under threat from changing climatic conditions as well as the demand for agricultural land from a growing population. He spoke about the need for a national strategy on ABS that could improve the livelihoods of local people through sustainable use and management of the resource at the local and regional levels (Taleb, 2014).

The argan ecosystem is an essential component of the ecological balance of the area and acts as a barrier against desertification (Lybbert et al., 2010). Clearing argan lands for intensive agriculture systems, overexploitation for wood and intensive grazing are posing an increasing threat to the existence of the argan woodland (Janick and Paull, *eds.*, 2008). The deep roots of argan trees are the most important stabilizing element in the arid ecosystem, and as the poor forest dwellers depend on the tree for grazing by their livestock, there is a persistent threat to the species. Particularly the goats are found browsing the argan trees heavily, climbing into its canopy. The trees are slow growing, and are not endowed with fast regeneration rates (Janick and Paull, 2008; Moussouris and Pierce, 2000). The general poverty and illiteracy in the region are relatively high, and it is a matter of great concern that the argan tree has defied domestication (Lybbert et al., 2011), which is a call for the conservation measures to be taken up for the species. It was observed (Lybbert et al., 2011) that the households involved in argan oil supply chain, with rising argan oil prices, acquired more goats as additional assets, and not other livestock, leading to a negative impact on forest health.

In 1998, under the Man and Biosphere programme, United Nations Educational, Scientific and Cultural Organization (UNESCO) classified the argan region of 2,568,780 hectares as a Biosphere Reserve[13] that includes the argan forest areas of 828,000 hectares (Guillaume and Charrouf, 2011). The argan tree is the focus not only for conservation but also for research and resource use in socio-economic development, and therefore the remaining forests need to be protected and assisted through reforestation efforts (Lybbert et al., 2010, 2011; Taleb, 2014). Due to this status, 2 per cent of the Argan Forest Biosphere Reserve has been protected from human activity.

The Argan Oil

The argan oil, extracted from the nuts of argan, is in great demand at global level today. It is well accepted as a nature-based product, being used for centuries in Morocco as cosmetic oil to maintain a fair complexion and to cure skin pimples and chicken pox pustules

---

[13] *Source*: www.unesco.org/mabdb/br/brdir/directory/biores.asp (Accessed in November 2014).

scars (Villareal et al., 2013). Some of the properties of argan oil, like its de-pigmenting effect due to the active ingredients present in the oil (fatty acids, tocopherols and carotenoids), have been reported, leading to its use for protection of the skin from sunlight. The oil as a natural food may reduce cardiovascular risk, as Moroccan researcher Anas Drissi and his fellow scientists found that regular consumption of the oil induces a lowering of low-density lipoprotein (LDL) cholesterol and has antioxidant properties. After investigating the effect of regular virgin argan oil consumption on lipid profile and antioxidant status with Moroccan subjects, it was found by the researchers (Drissi et al., 2004) that diet composition of argan oil consumers had a higher significant content of polyunsaturated fatty acids than that of non-consumers, and subjects consuming argan oil showed lower level of plasma LDL cholesterol compared with the non-consumers.

Villareal et al. (2013) performed the chemical analyses of the cosmetic argan oil used in the experiments to find the amount of the major fatty acids. They found that the argan oil contains oleic acid (46.1 per cent), linoleic acid (34.5 per cent), palmitic acid (12.2 per cent) and stearic acid (6.23 per cent), and its antioxidant properties provide additional benefit to the cells via the reduction of oxidative stress in the absence of melanin. The melanogenesis regulatory effect of argan oil was also evaluated by the researchers, and the results showed a dose-dependent decrease in melanin content, providing the scientific basis for the traditionally established benefits of argan oil and its therapeutic potential against hyper-pigmentation disorders after exposure to sunlight for a long time. Its anti-diabetic activity has been demonstrated in animals, and the main therapeutic properties of the oil are listed by Guillaume and Charrouf (2011) under the category of 'Beauty' as anti-aging, moisturizing, wound healer, anti-acne and anti-sebum (sebum is oily secretion of the sebaceous glands), and under the category of 'Edible' as hepato-protective, anti-diabetic, anti-proliferative, anti-obesity and choleretic.

The *Encyclopedia of Fruit and Nuts* (Janick and Paull, 2008) mentions that the oil contains 99 per cent of fatty acids, of which 80 per cent are unsaturated, with oleic acid and linoleic acid (polyunsaturated, omega-6) being the major components, and the oil is rich in vitamin E too. Recent studies in animals and humans have strengthened the belief in the health properties of the oil. The fatty acids have numerous therapeutic effects, and therapeutic doses to prevent metabolic diseases range from 15 to 30 gram of uncooked argan oil per day (Guillaume and Charrouf, 2011).

Well, it is quite apparent why the price of argan soared between 1999 and 2007 in high-value markets and sparked a bonanza of argan activity (Lybbert et al., 2011).

Women's Cooperatives

The production of argan oil is basically women's work. The kernel of argan fruits for the extraction of the oil are picked manually by women belonging to the local communities and processed locally. For centuries, the hard work of extracting oil from the nut of the argan tree has been performed by Berber women, the indigenous people of North Africa living in the argan region. For centuries, the Berber women have been using their TK to extract the argan oil. The plum-sized fruit contains one to three kernels with an oil content exceeding 50 per cent (Moussouris and Pierce, 2000). The average annual fruit yield is 8 kilogram per tree (Janick and Paull, 2008). Extraction of the oil, generally done by woman only, is a tedious work. The hard shells are broken with a stone and the kernel is extracted and chopped. After adding some warm water to make a rough paste to help extract the oil, the mixture is pressed in a small home-made mill made of large stones. Another method of extracting the oil involves separating the dried flesh to take out the nut; the seeds are then lightly roasted, ground and mixed with warm water followed by rinsing, which separates the floating oil. Working hard for about 20 hours, from an average weight of 34 kilogram of dried fruits, the women are able to extract just about 1 litre of argan oil (Glaser, 2010). The oil is extracted from roasted as well as unroasted kernels, depending upon the end use as the process results in varying properties of flavour. The argan fruit can be stored for many years without affecting the quality of oil ultimately extracted from the kernel, and the oil is extracted from the leaves too (L'Oréal Canada, undated).

In the early 1990s, a chemistry professor at University Mohammed V in Rabat, Morocco, Zoubida Charrouf, decided to conduct chemical research on the oil[14] and began organizing argan oil cooperatives for women in the argan forest region in order to protect women's interests and improve their status, and by the year 1996, she had created a network of several cooperatives called Targanine (Glaser, 2010). There are 500 women working in six production cooperatives, one extraction and oil facility and 15 preparation cooperatives. Shipping is organized through a commercial organization (Groupement d'intérêt économique) who represents 1 per cent of argan oil activities in Morocco (L'Oréal Canada, undated). Over the years, a practice has evolved in which women cooperatives are engaged for the extraction of the oil and its marketing. In this context, Lybbert et al. (2010) stated that two different cooperative models emerged for pursuing the objectives of empowering women and conservation of the threatened forests. The first was fuelled by Professor Zoubida Charrouf in the mid-1990s, when she began organizing argan oil cooperatives. The second effort was led by the German development agency German Corporation for International Cooperation (GTZ), which also supported the development of argan oil cooperatives for women, albeit of a different form. It is estimated that there are more than 100 women's cooperatives operating in Morocco servicing national and international companies (Taleb, 2014).

The argan oil attracted the media's gaze and has been showcased by many tourist publications and has dozens of websites dedicated to it, selling argan oils. New argan oil producers, distributors and cooperatives are springing up to pull the win-win story, many proactively linking up with cooperatives (Lybbert et al., 2010). In fact, women's cooperatives are prevalent today and function more like profit-making shops, posing as cooperatives, and they neither offer fitting benefits nor are certified by any governing body.

Benefit Sharing Arrangements

The rising of the argan oil price has increased the extraction in the region and, in addition, has resulted in its reduced traditional domestic consumption. Lybbert et al. (2011) found in their household

---

[14] *Source*: www.arganoilmorocco.com (Accessed in November 2014).

survey-based study that with rise in argan prices between 1999 and 2007, the average household stocks of argan fruits increased 10 times speculating enhanced profits in future, the average household oil production tripled and average domestic consumption halved. They also found that the locals were benefiting from the argan boom in ways that may improve women's welfare and alleviate persistent rural poverty. As reported, basically, it has made many women there the main breadwinners at home, encouraging other women to set up businesses of their own.[15] However, the study also shows that the households that previously only had seasonal usufruct rights to specific tracts of the forests are now enclosing these tracts to deny others access all year long. This is at the cost of excluding poor households with limited private productive resources from these forest commons. The poverty of these households may persist despite booming argan markets.

Unfortunately, it is found that argan oil commercialization is not turning into the net conservation incentives, and the argan forest is inviting long-term harm (Lybbert et al., 2010).

Corporate Social Responsibility

As has been mentioned in the beginning that the business of argan oil of Morocco is a unique case of benefit sharing arrangement, for it adopts the CSR approach, let this aspect be examined critically in the perspective of the women's cooperatives. It is established that the extraction of argan oil is basically women's work, in which the Berber women, using their TK, do the tedious work of oil extraction. It is also proven that with the women's cooperative model that emerged follows a process of pursuing the objectives of empowering women, so much so that it has made many Berber women the main breadwinners at home.

In this framework, some of the available references indicate the CSR approach of Laboratoire Serobiologiques (LS)[16] partnering with the multinational cosmetics company L'Oréal and Yamana, an NGO.

---

[15] *Source*: www.bbc.co.uk/news/business-16460127 (Accessed in November 2014).
[16] A France-based multinational supplier of cosmetic actives, which accentuates CSR as a key value for its success and lays emphasis on key innovations in cosmetology.

These three bodies initiated a tripartite partnership and involved the Targanine cooperatives of Morocco for the sustained supply of the oil and other related argan products. The processed products are incorporated into a number of cosmetics products marketed globally. The Yamana assumed the role of trainer and facilitator for the Moroccan cooperatives to ensure fulfilling the stakeholder expectations as well as sustaining the supply chain (Robinson and Defrenne, 2009).

Some of the most important aspects of this approach have been:

1. Purchase of products from Targanine cooperatives with the intention of providing employment, inculcating the sense of ownership and shared decision-making and, as the final result, empowering the Berber women;
2. Ensuring that the women from Targanine cooperatives are paid at the rate of per kilogram of argan nuts above market price; and
3. Of all the proceeds, approximately 50 per cent goes into the social fund, 25 per cent to the Targanine for maintenance costs, machinery, investment and management and 25 per cent to the cooperatives. The social fund is spent on eyeglasses, literacy programmes and basic hygiene-related products such as washing machines, pharmaceuticals and health insurance cover.

Commentary on the Argan Oil Case

There are few issues of conflicts and some riveting concerns. Firstly, replacing much of the traditional extraction methods of argan oil, a mechanized process of extraction was introduced by Professor Charrouf to ensure the quality of the product. Although the mechanized process extended the oil's shelf life from six months to two years, the women still crack the stones, because no machine can match their dexterity.[17] It was also observed, as Glaser (2010) states, that middlemen appeared on the scene, which obviously started disturbing the cooperative network and cracking the financial balance, because of which undoubtedly losers were the women. In the beginning, men were apprehensive about the success of the cooperatives and scoffed

---

[17] *Source*: www.arganoilmorocco.com (Accessed in November 2014).

at Professor Charrouf's plans, but seeing the economic benefit of the cooperatives, they began encouraging the project. Concurrently, as has been alluded to earlier, unwelcome happenings also are creeping in as some of the women's cooperatives are functioning more like profit-making shops, only posing as cooperatives, and they neither offer fitting benefits nor are certified by any governing body.

The cooperatives created new jobs paying a decent wage, and Professor Charrouf offered classes in reading and arithmetic.[18] On the conservation front, each of Charrouf's workers began planting 10 new argan trees a year, and significantly, the villagers have become much more reluctant to let their goats graze in the trees. If we look at the scenario in the perspective of ABS laws, it is observed that no strict rules are being followed concerning the rights, whether relating to the cultural or IPR, and there is no clear regulatory status over patent disclosure requirements or agreements between the different stakeholders (Glaser, 2010). The multinational company L'Oréal exploits the argan oil's value commercially, and although on their website they report about their CSR and mention their support for women scientists, there is no specific collaboration or cooperative named by L'Oréal (Glaser, 2010).

In this context, it may only be conjectured that the LS is not liable to conform to ABS regulations in the tripartite arrangement between the LS, L'Oreal and Yamana, as Morocco does not have an ABS law today, although the country is a signatory to the CBD and has ratified the Nagoya Protocol. But so long as the prevalent benefit sharing arrangement with CSR approach exists, the arrangement should be able to sail through even after the country adopts such laws.

The Options Available

It is perceived from these case studies that the benefits accruing from the local biodiversity are 'shared' between the industrial groups and the indigenous and local communities. Is this sharing equitable? The bioprospecting contracts between companies and communities are

[18] *Source*: www.arganoilmorocco.com (Accessed in November 2014).

looked at with certain apprehension and are seen as having a tendency to skew the flow of benefits. As an exposé of the GRAIN (based in Barcelona, Spain) and the Gaia Foundation, London (GRAIN, 2000), points out, while bilateral contracts between bioprospectors and local communities can in specific cases help generate additional income and other benefits for local communities, by and large they show that the vast majority of the benefits derived from biodiversity continues to be captured by industrial interests—in most cases way over 95 per cent—rather than by local communities or developing countries.

The skewness in benefit sharing as indicated above is a natural consequence of the state of affairs wherein, firstly, the developing countries hold immeasurable biological diversity, representing largely untapped source of genetic resources that are primary source of livelihood for the people living inside and on the fringes of forests, and secondly, there is a clear indication that the general flow of trade in genetic resources is directed mostly from the resource-rich and technology-poor 'South' to the biodiversity-poor and technology-rich 'North', which is not recouped to scale. This ironically happens despite the new market preferences enhancing the demand in the international trade for the natural products in general. Incidentally, the products of biodiversity of the tropics lack appropriate statistical information, and the researchers experience constraints with regard to the study of benefit sharing aspects nationally as well as internationally. Moreover, the biodiversity-rich countries of the 'South' fear that its genetic resources and the TK associated with them remain unprotected under the new IPR regime launched by the WTO.

The developing world has been pursuing for review of the Agreement on TRIPS for incorporating suitable amendments with the support of the provisions laid down in the CBD. India's Biological Diversity Act of 2002 appears to be a significant legislation in respect that no international agreement has as yet been arrived at as to how to put into effect the relevant CBD provisions, and the Act should help direct proper flows of benefits of commercial uses of biodiversity to holders of TK. The Act is being reviewed and studied by many, and many other countries are presently in the process of finalizing similar benefit sharing enactments. It is noteworthy that the implementation of the provisions of India's Biological Diversity Act of 2002, which is quite elaborate and comprises the base provisions for ABS principles, is an all-inclusive law. As reported in various forums and documents, the Act is not touching the ground far and near in different states of the country with the pace it should have been doing so, and a lot many efforts on the part of the NBA have been made, but many more action points are still anticipated, including the circulation of vernacular versions of the guidelines for ABS on the lines of the Nagoya Protocol, which India was one among the first countries to ratify.

In a nutshell, it is established that the demand for genetic material from the 'South' is likely to continue and the 'South' has to find out how to benefit from the present IPR regime. Several questions remain unanswered: What are the possible mechanisms that can ensure equitable commercialization of the local biodiversity? Could a mechanism be devised to enable the indigenous and local communities to be the co-inventors of the new products and processes? And lastly, is it possible to evolve strategies that could remove all the anomalies so that an equitable sharing of benefits could be ensured? All these and many other related questions need to be examined.

The case studies reviewed and studied in depth in this chapter make it evident that even though the limitations exist as regards the protection of the TK associated with the genetic resources in the global IPR regime today, and the demand for nature-based products are ever increasing, there are alternatives to regulate access to the genetic resources. A set of options are available for an equitable sharing of benefits accruing from the biological resources of the tropical countries between the indigenous and local communities and their pharmaceutical partners in bioprospecting partnership contracts. It is also evident that the trust fund mechanism is the best and only way to ensure an incessant flow of the funds involved in any biodiversity partnership programme. Based on the review of these case studies, an attempt is made in Chapter 7 to work out a mechanism for bioprospecting partnership programmes that may prove beneficial to all the on-site and off-site stakeholders who are involved in these partnerships.

7

Forging Bioprospecting Partnership Contracts: Mechanism for Equitable Benefit Sharing

The huge attention drawn to the biological diversity all over the world during the last 25 years has been appraised in the previous chapters, and the dependence of the indigenous and local communities in tropical countries on these products has been discussed vis-à-vis the role and the rights of the people in sustainable management of these resources. The anomalies in the sharing of benefits accruing from these resources were also discussed, and in this chapter, the focus will be on forging suitable partnership agreements that are free from anomalies. In this perspective, it is well recognized that the well-being and the socio-economic development of the indigenous and local communities are interwoven with the sustainable management of genetic resources as the diversity of biological resources is of critical importance for meeting their food, health and livelihood needs. It is also established that there is a growing demand in the world market for natural pharmaceutical, cosmetic and other products, that the general flow of trade in various floral as well as faunal species is directed mostly from the resource-rich and technology-poor 'South' to the biodiversity-poor and technology-rich 'North' and that the developing countries are in the need of new and additional financial resources and appropriate access to relevant technologies. In no way it is simple to decipher about the quantity, the quality or names of the species traded in actual sense within the national boundary of a tropical country or at the international level; the trade in these resources is so secretive, skewed and unorganized.

Along with the issues related to the commercial use of biological diversity that is under focus, the access to biological resources with or without the consent of the indigenous and local communities who harness these resources and the issue of sharing of the benefits accruing from these very resources with the resource providers, another important subject that draws attention is the TK in possession of the indigenous and local communities. The TK plays a vital role in the socio-economic and cultural milieu in the developing and the least developed countries of the tropics. An argument is frequently set forth that the TK and genetic resources that they are associated with put the countries that are biodiversity-rich, though technology-poor, at the comparative advantage. The wealth of the biodiversity and the associated TK undoubtedly makes available an opportunity to those countries to participate more effectively in the global market where drug development is a big business and the demand for natural products is growing rapidly.

Yet, this advantage appears to be devious, for the TK per se is stated to be in the public domain and its protection in the present-day IPR regime is unachievable. This argument, however, has been well contested, and well-thought-out justifications are now spoken out to buy legal protection for the TK and to allow the coveted benefits to flow to the biodiversity-rich countries. These countries find that the arrangements provided under the CBD or the Agreement on TRIPS, within the larger umbrella of the WTO, are not adequate or appropriate for the protection of TK and the commercial use of biodiversity is not as beneficial to the indigenous and local communities as it should be.

Despite these limitations, it is feasible to have a workable mechanism for ABS as has been observed in the case studies discussed at length in the previous chapter. Drawing clues from the diverse partnership programmes discussed in these case studies, an attempt has been made in this chapter that is more of an exercise to work out an alternative so that the commercial use of biodiversity could be rendered beneficial to the indigenous and local communities. As a part of this exercise, a flow diagram (Figure 7.1) has been conceived in order to work out the mechanism for ABS. Concurrently, another flow diagram shall be presented later to show the trust fund mechanism that is increasingly being assumed as an effective medium to control the channelling of benefits from the commercial use of biodiversity.

Figure 7.1:
Mechanism for access and benefit sharing

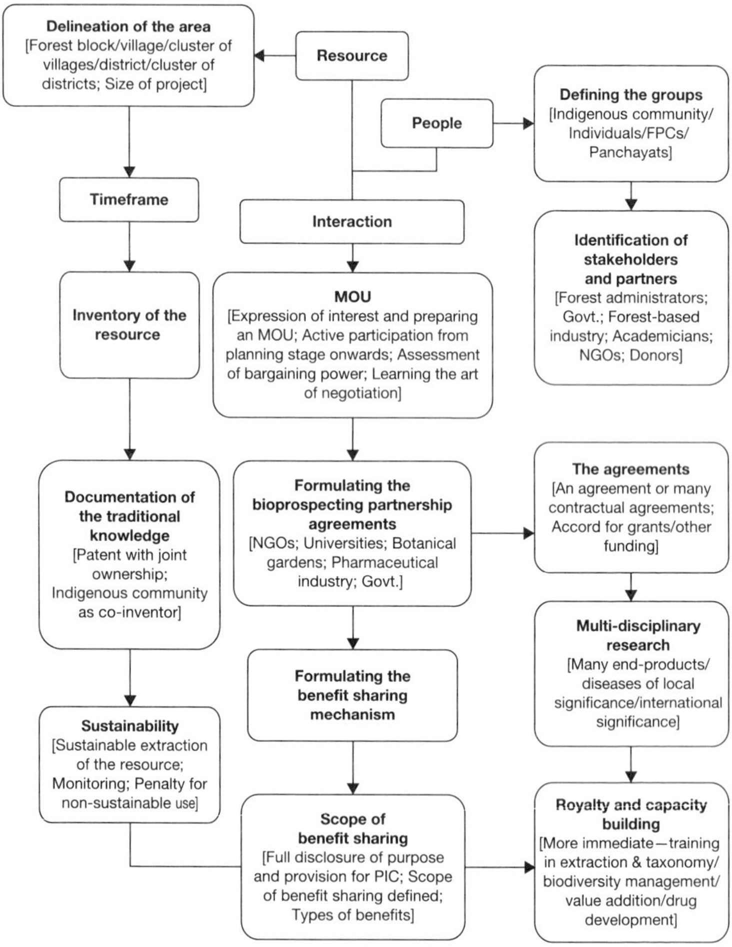

FPC: *Forest Protection Committee;* **Panchayat:** *Local self-governments headed by a 'sarpanch' at the village level in India.*

Mechanism of Benefit Sharing

It was seen in the case studies discussed in Chapter 6 that different ground realities of various partnership programmes were given different kind of efforts to devise the mechanism for the sharing of benefits from the biological resources. An attempt is now made to generalize the mechanism for sharing of benefits, keeping in view how to circumvent the implications of the WTO regime or, more specifically, the Agreement on TRIPS. Before deliberating on this mechanism, it can be reassured that there is no way except sharing of the biological resources of the 'South' and strive for the transfer of relevant technologies from the 'North' to the 'South' without which either the 'South' might be deprived of the innovations needed for bioprospecting or it may take too long, on its own, to draw the benefits of bioprospecting.

The reason why it is reassured that there is no way except sharing of the biological resources lies in the inference drawn earlier that the commercial enterprises could, in any way, patent products and processes derived from TK by making small 'improvements' (such as isolating an active component and patenting the extraction process) without any of the benefits accruing to the custodians of the TK. The provisions under Article 27 (*Patentable Subject Matter*) of the Agreement on TRIPS pose a threat to the biological resource and associated TK belonging to the 'South', and a mechanism needs to be evolved to thwart this risk often defined as 'biopiracy' until the Agreement on TRIPS accommodates one of the main objectives of the CBD—the fair and equitable sharing of the benefits from the use of genetic resources—after review and modifications in its respective provisions to suit the needs of the developing and the least developed countries of the tropical regions.

Similarly, another fact may be reaffirmed that there is no way except to look for a mechanism for regulating the access and simultaneously reassuring equitable sharing of benefits that could accrue from the commercial use of these resources. The reason is, restricting access to the tropical genetic resources might put a ceiling on innovation, and unregulated access could trigger their destruction and ultimate extinction that would benefit none. Still an unexplored wealth of genetic resources exists in the tropical countries and the humankind is undeniably in the need of more and more nature-based pharmaceuticals, cosmetics, food supplements and so on. The fight over these issues cannot and should not be afforded for very long years.

Now, parallel to the processes going on at the negotiation tables all over the world, the way to achieve that is to continue the commercial use of the biodiversity of the world through an equitable sharing of the benefits drawn. The way is shown by the case studies on hand, and a further look on this genesis leads to the formation of basic structure of a bioprospecting partnership that may comprise the following elements:

1. Involvement of various stakeholders from the planning stage, strengthening of their bargaining power and training in the art of negotiation;
2. Substantive material and monetary benefits to various on-site stakeholders;
3. Capacity building of indigenous and local communities along with technology transfer;
4. A comprehensive strategy for the protection of TK and joint ownership of the IPR;
5. Sustainable management of genetic resources;
6. Institution building (trust funds, etc.) for the purpose of benefit sharing; and
7. Chart out the most trustworthy bioprospecting partnership programmes.

The case studies deliberated upon in this book make it clear that the size of a partnership programme depends totally on local factors such as the extent and variety of the resource base, the size of the community involved, the scope of bioprospecting and the socio-economic conditions of the indigenous and local communities, and of course the role the funding agencies are willing to assume. Obviously, a bigger partnership programme should be flexible enough to accommodate variations in the socio-economic and cultural conditions of different communities through formulation of locally chalked out sub-programmes. The mechanism proposed on the basis of the above genesis may be given a shape as shown in the flow diagram (Figure 7.1), and its elements are described in the following sections.

A representative kind of mechanism that may be adopted under a specific partnership programme should commence with the delineation of the area of a particular project and an estimation of its size on financial scale, followed by defining the groups (the identification of the on-site and off-site stakeholders). Once these parameters are fixed tentatively (to be finalized at the time of formulating the partnership agreements), a broad framework of the programme and its time frame is drawn. The biological resources of the region to be accessed for commercial use must be inventoried simultaneously, for which suitable sub-programme needs to be drawn.

The final step will be taken up with formulating the different partnership agreements (mostly multilateral contracts) with appropriate inbuilt benefit sharing arrangements that would include all or an assortment of the upfront, milestone and advance payments, royalty sharing, capacity building, training and so on. The steps involved are sequenced in the following sections.

Delineation of the Area and the Size of the Project

There have been wide-ranging bioprospecting partnership programmes, extending over a country, and also some very compact projects encompassing merely a cluster of villages or spreading over a district taken up in different parts of the world. For example, the ICBG project, launched in 1992, was a large programme practising an experimental, multidisciplinary, relatively large-scale approach to drug discovery. Several of the ICBG projects have been completed, and many new projects have been undertaken the world over in South, Central and North American countries and also in Africa, Asia and Oceania. The ICBG projects are in progress presently in Costa Rica, Fiji, Indonesia, Madagascar, Panama, Papua New Guinea and the Philippines, and in some of these countries, even the second phases have been launched.

The project comprises diverse public and private institutions including universities, environmental organizations and pharmaceutical companies that are collaborating on multidisciplinary projects on the interdependent issues of drug discovery, biodiversity conservation and sustainable economic growth. For a five-year cycle of the programme, six awards of approximately $500,000 to 600,000 per year were made in 1998. The project aims at building research capacity in more than 20 different institutions and training hundreds of individuals. To date, thousands of species of plants, animals and fungi have been collected to examine biological activity in 19 different therapeutic areas. Numerous publications in chemistry, biodiversity policy, conservation and ethnobiology have been brought out by the biodiversity researchers.

Another example of a large programme is the FIRD-TM that was launched in Nigeria in 1997 by a multiethnic international NGO, the BDCP, to receive and channel benefits from many contributors, including the ICBG. The Fund has a 21-page constitution and an independent board composed of leaders of traditional healers' associations, senior government officials, multiethnic representatives of village councils and technical experts from scientific institutions. It has three principal organs: the board of trustees, the advisory board and the board of management, each with defined role and function, and is designed as a long-term national fund, unlike the FPF of Suriname that operates at the community level. At its inauguration, a kitty of $40,000 was donated to FIRD-TM for a pilot project, to test the efficiency and efficacy of the trust fund process. Town associations, village heads and professional guilds of healers are empowered to make decisions regarding use of the funds for projects following the criteria of promoting conservation of biodiversity and drug development, as well as the socio-economic development of rural cultures.

Another figuratively large collaboration has been between the INBio, a nonprofit, semi-public organization established by Costa Rica's MINAE and Merck & Co., a US pharmaceutical corporation (now a classic case of its kind, often referred to as the INBio–Merck collaboration), with an advance payment of $1,135,000 to INBio in 1991. To date, INBio's biodiversity prospecting agreements have yielded more than $390,000 to MINAE; $710,000 to conservation areas; $710,000 to public universities; and $740,000 to cover activities within INBio, primarily the national biodiversity inventory (Laird and Lisinge, 2002).

Compared to these, the Kani experiment of the JNTBGRI in Kerala (India) deals only with one plant resource, Arogyapacha (*Trichopus zeylanicus*), that is endemic to the region of the Western Ghats in the Thiruvananthapuram district of Kerala. The team of scientists isolated 12 active compounds from Arogyapacha, developed the drug *Jeevani* and filed two patent applications on the drug. The technology was sold to AVP for a licence fee of ₹1,000,000 plus a royalty share of 2 per cent. The trust (KKSK Trust) established for benefit sharing with the Kanis in 1997 covered over 700 families.

Another single species-based benefit sharing case comes from South Africa where the pulp of Hoodia (*Hoodia gordonii*) plants are eaten by people of the San community as an appetite suppressant that gives energy when they go to hunt or to gather products from forests. The TK belongs to the San community of Kalahari Desert. South Arica's CSIR recognized the TK of the people and worked on it to identify the properties of a specific molecule that was termed as 'P57'. The CSIR saw a scope of commercializing this molecule as an anti-obesity drug, the royalty from which could be shared with the San community. The CSIR got its first patent in 1996 and another from the USPTO in 1999. Though initially the patent aspect involved legal disputes between CSIR and the San people, the issues were resolved later when the San people formed the South African San Council and worked with the Cape Town-based San Institute of South Africa, an NGO, and the WIMSA, resulting into the signing of an MOU in 2002 that led to the provisions for milestone payments as well as provisions for sharing in royalty. This case is quite interesting in the sense that though formulation of the Biodiversity Law was going on and another Act—the National Environmental Management Act 1998—was already in existence, it was a government agency (CSIR) that was depriving the local community initially of the benefits that was due to them, but later it had to yield to the community's San Council for benefit sharing.

The outline of the sizes of the programmes described above should help in settling on the extent of the project area as well as the financial extent desired in a bioprospecting programme. It would then be the prerogative of the groups involved to decide the basic objectives to be achieved, depending on the financial capability of the probable participants, tailoring it to the socio-economic conditions and cultural traditions of the local community. It may be emphasized that smaller organizations can handle smaller and localized projects, while large-size programmes should formulate multiple projects governed by multiple partnership agreements and should have enough flexibility for accommodating regional, cultural and ecological variations. It is also noteworthy that involvement of a government agency is advocated in benefit sharing arrangements to ensure authenticity and trustworthiness, as is discussed further in this chapter, yet the way the CSIR in South Africa behaved could set out that it would not be enough that a government agency is recognized as one of the stakeholders and everything would be set in order. The government institutions also ought to behave in a pragmatic way.

In addition, it is essential to have the project size well defined and the programme area also well delineated, specifying the indigenous and local communities that the project would comprise, for if the community is not well defined geographically or otherwise, and if there is ambiguity as regards the principal beneficiaries (whether they should be individuals, or groups who actively participate or everyone in the community), returning benefits to communities in the benefit sharing process can be extraordinarily complicated in practice (Rosenthal, 1997a). Concurrently, it is very important to ensure that the geographical region is well delineated to cover the entire ecological region wherein the biological resources that are to be put under commercial use originate. If the biological resources that are available in a particular ecological region are of similar nature and carry equal commercial value and future scope of development, and are found to inhabit different ecological regions, then this issue deserves to be paid special attention, particularly in the perspective of TK that may be prevalent across the different ecological regions. This approach may curtail the demand arising in future from other benefit claimers of similar ecological regions.

Defining the Groups

Indigenous and local communities in the developing countries in general have continued access to their local natural resources as a right. They are marked with 'high importance' with respect to the NWFP resource in almost all the exercises of stakeholder categorizations and the approach of giving them priority is fairly relevant. It is noteworthy that though the governance decisions are taken at the regional or even national level, the status of resource and its management at the local level are essentially influenced by the local factors (Kameswari, 2002). The local factors include the socio-economic, cultural and ecological factors, local traditions and especially the characteristic of the peoples' involvement in common pool resource. It is also argued (Kameswari, 2002) that prescriptive policies regarding the functioning of local-level institutions adopted by the state have failed to involve the user group in resource management. It is in this light that an appropriate emphasis ought to be there particularly in the case of local traditional healers as to how they are included in various bioprospecting partnerships, whose wisdom was always the key to most of the new drug discoveries. Similarly, the sustainability is another issue that depends predominantly on the local traditional practices of resource exploitation. A few examples are cited in the following paragraphs to illustrate how the non-involvement of certain groups led to failure of or delay in some partnership programmes.

The herbal drug, *Jeevani*, was discovered only on the information given by two Kani guides to the scientists of the AICEP on a botanical expedition in the Western Ghats. The Kani tribes were assured by JNTBGRI, the research institution, for a share of 50 per cent of the fee and royalties that would come from *Jeevani*. However, all the Kani settlements were not considered for benefit sharing, and the role of the *Plathis* (tribal physicians among the Kanis) also was not recognized in the benefit sharing arrangements. A disgruntled group of Kani medicine men wrote to the Chief Minister of Kerala in 1995, objecting to the sale of their knowledge to a private company.

It is also stated that the process of trust formation in the Kani's case could have been more participative and the rights of informants (the two Kani guides whose information led to the discovery of *Jeevani*) and that of the community would have been distinguished in the benefit sharing arrangements. The informants were the first to receive payment from the amount deposited in the trust fund, while they could have been paid from the resources received by JNTBGRI, and the impression was that the trust was targeting to benefit only a few community members (Gupta, 2002b). Besides the criticism of mode of sharing of benefits, the participants of the programme also showed lack of awareness about the functioning of the trust.

This example depicts that consultation with the local community, aiming to ensure their effective involvement, is a complex issue and needs careful dealing. In a study of the experimenting of Karnataka and Kerala states in India with formulating the PBRs, concern was shown by some people at the lack of direct consultation with local and tribal communities, and it was stated that though NGOs helped to reflect the interests of local communities in this experiment, they could not be equated with elected representatives (Anuradha et al., 2001). The plus point with the involvement of local communities is that it inculcates a lot of encouragement, and building on accepted institutions of TK-holders can be an important tool to structure their participation and ensure the acceptance of the communities. A heartening response of the local communities was observed in the FPF in Suriname where the Maroons seemed to understand and accept the broader ultimate objectives of the ICBG project. It clearly indicates that though the process of getting the local communities involved in the negotiation process, more so if diverse clans or groups of local communities coexist in proximity of the region under the project, they need to be involved keeping in mind that they continue to remain the most intrinsically associated on-site stakeholders.

However, not all groups of the local population respond positively. The Saramakas in the Suriname were quick to appreciate the immediate, here-and-now benefits of the project's work in their territory, although the rank-and-file Saramakas did not understand the specific purpose of the project. Under such situations, the institutions of local self-government like village Panchayats and village forest or forest protection committees (FPCs) present good prospects in securing community participation. It is also true that, in practice, securing effective community participation, though a complex job to be done, would require the dissemination of information in local languages and seminars and discussions at local level to elicit responses (Anuradha et al., 2001). Probably that could be the best option to define the groups that are aimed to be associated and involved to the fullest extent.

A moot question is also raised as to who could represent the indigenous and local community in the best possible way in a given countryside? The WIPO Publication No. 2 of the series dealing with intellectual property and genetic resources, TK and traditional cultural expressions/folklore suggests, in the context of the TK often being held collectively by communities rather than by individual owners (which is often cited as a drawback also in protecting TK), that it is possible to form associations, community corporations or similar legal bodies to act on behalf of the community (WIPO, 2004a). This suggestion is well proven when we review the various case studies on benefit sharing. Well again, while eliciting the support of various groups at local level, it is also essential to eliminate prejudices and offset all kinds of biases, having a team of multidisciplinary members with the ability to define and distinguish the technical, social and economic constraints of the traditional cultures. The well-known method of 'triangulating'[1] also may be tried to cross-check answers as it involves comparing and complementing the information gathered from different sources through different methods.

Evidently, it may be stated that the exercise of identifying and defining the various groups who matter most as the on-site stakeholders ought to be taken up much in advance, as the programme objectives and major goals would be closely linked to these groups. The identification of stakeholders involved and deciding about the actual partners in various agreements are the next important tasks, to be followed by giving shape to the larger framework and major goals of the partnership programme.

Identification and Analysis of the Stakeholders

Before actually entering into the process of multilateral agreements, the most imperative arena is the identification of stakeholders and choosing the major partners from the off-site stakeholders. Therefore, it is essential to qualify the different categories of stakeholders—some assuming central role by virtue of their capacity and few others whose very existence depends on the resource base involved. An outline of the identification process of different stakeholders ranging from the local level to the international level and an analysis about their influences on each other is recounted below.

When we deliberate upon the commercial use of biodiversity and the sharing of benefits accruing from these resources, we find that several stakeholders are involved in the process, namely, members of the indigenous and local communities (particularly the NWFP gatherers), traditional healers, local traders and other middlemen, forest workers, local bodies and many other interest groups. These groups are in general termed as on-site stakeholders, and conversely, there are researchers (taxonomists, ethnobotanists, etc.), exporters, manufacturers, research institutions, NGOs, funding agencies and so on who are termed as off-site stakeholders. It would be proper therefore to

---

[1] The triangulation means the confirmation of findings by different methods, which improves the overall validity of results, and makes the study of greater use to the constituencies to which it was intended to be addressed (Maxwell, 1998).

apply stakeholder analysis and influence mapping of various stakeholders involved in the commercialization of the components of biological diversity. This exercise is essential chiefly when multilateral and long-term bioprospecting partnership programmes have to be worked out. Such programmes are quite complex, requiring the process of weighing many mega-influences and engaging numerous institutions. There may be a requirement of different sub-programmes also, to be designed to match local variations with specific agreements for each sub-programme.

Involvement of a range of stakeholders begins from the planning stage, the exercises in benefit sharing often involving capacity building of stakeholder groups along with technology transfer as well as institution building for the purpose of benefit sharing. At the outset, it is realized that biological resources, especially the endemic species if there are any, over which the indigenous and local communities have usufructs, are multi-stakeholder, and when we begin to deliberate on this, some basic questions arise as to how the on-site and off-site stakeholders influence the resource as well as each other, are there any conflicts involved and could there be a single formula for equitable distribution of benefits among all the categories of the stakeholders?

Characteristically, different stakeholders have different goals and are often in conflict with one another. This is because they differ in their livelihood goals, resource use and access, have their own vision, values and priorities, and also they find it difficult to understand information communicated by other stakeholders (Boutelle, 2004; Lawrence, 2003). These differences in interests have been scrutinized and discussed in more detail in the case studies on partnership programmes in Chapter 6. Briefly it can be stated here that not only different groups of indigenous and local communities queue up with their perception for access to the resource or the benefits accruing from them, but even if the government sector is not involved from the very beginning in such programmes, there may be a culmination in which the outcome is hindered. It has been seen how the benefit sharing programme in the case of Kani tribe in Kerala (India) was stalled at much later stage when the *Jeevani* drug had reached the market after partnership programme took off between the research institute, the local community and the industry (AVP). The state forest department, which was not included as one of the stakeholders, brought in the fear that the endemic species 'Arogyapacha', which was the basic raw material for *Jeevani*, could vanish from the forests because of over-exploitation. And apart from the sustainability issue of the resource, another issue that came up was the claim of certain clans of the Kani tribe who alleged that they should also have been taken in. Even in the much talked about partnership programme of INBio–Merck, it is remarked that the people were never involved in the negotiations and the royalty-sharing agreements, which had clauses that are still a secret. John Eberlee in an IDRC report (Eberlee, 2000) evidently mentions that the royalty rates in the INBio–Merck biodiversity prospecting agreement were held strictly confidential, and the indigenous and local communities did not share in the economic benefits to any great extent.

Thus, it is quite obvious that suitable analysis of the stakeholders has to be an important component of any bioprospecting or benefit sharing programme, for different stakeholders have varying interests and influences in any partnership programme and the sharing of benefits. Bryson (2004) may be quoted here about the significance of stakeholder analysis: All it takes to do the stakeholder identification and analysis is some time and effort—an expenditure of resources that typically is miniscule when compared with the opportunity costs of less than adequate performance, or even disaster, that typically follow in the wake of failing to attend to key stakeholders, their interests and their information.

For conducting the stakeholder analysis, the potential stakeholders should be listed and classified broadly as the 'support' group or 'opposition' group depending upon their response to the programme in the wake of their interests that are likely to be taken care of or are likely to be affected adversely. They may be ranked as per their importance also, the evaluation of which is a very important step. For this purpose, they have been classified by FAO (undated) in each phase of the national forest programmes (NFP) process into the following four groups of importance/power:

1. High Importance/Low Power [forest owners (private-small), farmers, community organizations, local government, national NGOs, small-scale forest industry, indigenous people, forest workers]
2. High Importance/High Power [academics and research, forest owners (private-large), national governments/national forest

authorities, large forest industry, legislators, forest administrations, forest workers]
3. Low Importance/Low Power [other users, academics and research (developing country)]
4. Low Importance/High Power [donors, international NGOs (IUCN, World Wide Fund for Nature [WWF]), other ministries, international organizations]

The above grouping of importance of stakeholders is helpful in estimating the most genuine importance of each. This estimation is significant and will be more comprehensible if we refer to the case studies discussed in Chapter 6. For example, with reference to the FIRD-TM case from Nigeria, questions such as evaluating the interests of the groups such as traditional healers, the local communities and several unregistered healers remained unanswered because of lack of appropriate estimation of the role and importance of each faction. With reference to other case studies, it is remarked that the benefits accruing from the local biodiversity are not shared equitably between the industrial groups and the indigenous and local communities, and the contractual programmes are having a tendency to skew the flow of benefits in complete ignorance of the importance and influences of various groups of stakeholders. An appropriate balancing act could surely make it possible that the major benefits derived from biodiversity do not continue to be captured by industrial interests and the local communities, as well as the developing countries, which are the ultimate resource provider, also get an equitable share that they deserve.

In addition to the exercise of classifying various stakeholders as specified by FAO described above, a very useful tool is available that has been propounded by the CIFOR (1999) for putting the various stakeholders on a linear scale of the importance of the resource. In this method, a simple scoring technique is expressed for determining whose well-being at the local level must be taken as the central part of sustainable forest management. The method is presented in the Criteria and Indicators Toolbox Series of the CIFOR as a tool—the 'Who Counts Matrix'—for differentiating the main actors in forestry practices from other stakeholders. The matrix was worked out based on four major groups of stakeholders—local community, forest workers, timber company and local government. A marking is suggested in the matrix

Table 7.1:
Sample 'who counts matrix'

	Local Community	*Forest Workers*	*Timber Company*	*Local Government*
Proximity				
Pre-existing Rights				
Dependency				
Poverty				
Local Knowledge				
Culture/Forest Link				
Power Deficit				
Mean Value				

Source: CIFOR (1999).

depending upon the proximity of the stakeholders to the forests, their pre-existing rights, dependency, poverty, local knowledge, culture/forest link and power deficit. Originally formulated in 1995 for differentiating forest actors (people whose well-being and forest management are intimately intertwined) from other stakeholders, this matrix comprises seven dimensions by which forest actors can be differentiated from other stakeholders. A simple scoring technique is taken up for determining whose well-being must form an integral part of sustainable forest management in a given locale. The CIFOR has used it repeatedly in Indonesia, Côte d'Ivoire, Brazil, Cameroon, Trinidad and the USA. A sample 'Who Counts Matrix' is given in Table 7.1.

The left-hand column in the matrix lists the dimensions and the top row lists the stakeholders. An assessor scores the relevant stakeholder on the degree to which each dimension generally applies to that stakeholder (1 = high, 2 = medium, 3 = low and 'var.' = variable). The mean score for each column (excluding 'variable' scores) is computed across the bottom of each table. The stakeholders have been arranged so that the mean scores increase as we move to the right of the matrices. A reasonable cut-off point for defining forest actors seems to be a score of <2.

Results from the trials of this tool, tested by many, show that the method is handy and useful, and adaptable to local conditions. It is quick, easy and has yielded results consistent with the principles of the criteria and indicators for sustainable forest management. While

Table 7.2:
Influence mapping of various stakeholders

1	Local communities		High Importance/Low Power
2	Government bodies		
	(a) Local		High Importance/Low Power
	(b) National		High Importance/High Power
3	Private sector enterprises		
	(a) Small scale		High Importance/Low Power
	(b) Large scale		High Importance/High Power
4	Nonprofit intermediaries		
	(a) National NGOs		High Importance/Low Power
	(b) International NGOs		Low Importance/High Power
	(c) Academics and research (Developing country)		Low Importance/Low Power
	(d) Academics and research (Developed country)		High Importance/High Power
5	Consumers		High Importance/Low Power

the 'Who counts matrix' helps in assigning value to various categories of stakeholders, the technique of 'influence mapping' propounded by Walter et al. (2004) is used to reveal what affects the decision-making process, and a general list of stakeholders may be divided into various mappings as per need. For applying this method to the benefit sharing arrangements, the stakeholders are categorized broadly as local communities, government bodies, private sector enterprises, nonprofit intermediaries and consumers.

In the influence mapping exercise, any benefit sharing arrangement that has to weigh the influences of the main stakeholders would essentially have these five constituents. On arranging this categorization into the importance of the resource to the stakeholders' interests and the power they have to influence the management of the resource, a matrix appears as given in Table 7.2.

Almost all the exercises of categorization of stakeholders mark the indigenous and local communities with 'high importance/low power', and it is advocated to ideally aim them to a 'high importance/high power' status. The approach of giving priority to these on-site stakeholders in resource management at local level is fairly relevant because of an anomaly that the governance decisions are taken at the regional or even national level, whereas peoples' involvement in common pool resource institutions is essentially influenced by local factors. This observation is validated by a study taken up in Madhya Pradesh (India) by Kameswari (2002) on the management of forest lands by local user groups. She finds that the people have apathy and apparent lack of interest in the common pool resource management institution because of the feeling of helplessness and powerlessness among the resource users, and these feelings emerge from the fact that the actual users of the resource and members of the local-level institution have very little say in its functioning.

It would also be fitting to exercise a strengths, weaknesses, opportunities and threats (SWOT) analysis of the on-site stakeholders involved in the collection, processing and marketing of genetic resources for a better and broader understanding of their influences over the trade and benefit sharing issues. The outcome of the enumeration of their SWOT is described as follows:

Strengths:
1. The TK that is held by these stakeholders;
2. Their vicinity to forests (as the habitat of the indigenous and local communities coincide geographically with the forest cover of the country, and they meet out their livelihood and health-care needs from forests only); and
3. The nature-based life and culture (for it helps in the evolution of new and additional TK and also in the conservation of biological diversity, a repository of future natural products and drugs).

Weaknesses:
1. Vulnerability of TK that they hold, as it is not possible to prevent its misappropriation with the help of existing IPR laws; even more so when it is observed that there are lesser members of the younger generation of traditional healers who have willingness to carry on the practice.
2. Lack of right to the resources and lack of awareness, that leads to imbalances in benefit sharing; and
3. Non-involvement in the decision-making process, despite the role the biological resources play in meeting their basic livelihood and health-care needs.

Opportunities:
1. Biological prospecting from the natural resources might offer maximum return to the holders of TK if successful research programmes are carried out; and
2. The skill in the identification of species and the extraction processes, if upgraded, and value addition in the forest products might enhance the benefits drawn from these resources.

Threats:
1. Socio-economic dependence on biological resources, as often higher market prices of the resources compel the local communities to deprive themselves from self-consumption of such products; and
2. Unsustainable exploitation of the biological resources that are in high demand face the threat of overexploitation and ultimate extinction.

In a nutshell, what may be stated is that the interests of the on-site stakeholders, that is, the indigenous people and the community organizations (village forest/FPCs, joint forest management committees and the local self-government institutions like the Panchayats), should be emphasized, and other stakeholders seemingly significant in benefit sharing deliberations would be: national/state governments—forest authorities, other ministries, small-scale forest industry, large forest industry, academics and research (developing country), national NGOs, international NGOs (IUCN, WWF), donors and intergovernmental organizations.

There are a range of other investigation procedures to analyse and identify various stakeholders that are taught in training programmes around the world and have clear-cut relevance for such analyses. The rapid rural appraisal (RRA) methodology, which was found successful as an extension technique promoted by the Consultative Group on International Agricultural Research Centers (CGIAR), is effective when the baseline data and trained enumerators are inadequate. Likewise, the PRA approach to investigate the opinions of local people for planning and management of local resources, which is a semi-structured activity quite suitable to be used by multidisciplinary teams, also may be found to be very instrumental.

These approaches have inbuilt advantages. While the matrix scoring method is more helpful as it gives linear scale for measuring influences of various stakeholders, the RRA, PRA or other related technique known as the participatory learning and action (PLA) need to be applied before entering into bioprospecting partnerships. The techniques are based on group dynamics including feedback sessions, sampling including social mapping and semi-structured interviews and of course the oft-used Venn diagrams. These are helpful to (a) search relevant and new information, (b) monitor and evaluate marketing activities including price efficiencies, (c) identify and assess the capabilities of local institutions, (d) understand the resource use and dependency patterns of local communities, (e) prepare the inventory of the biological resources and (f) formulate new hypotheses even.

Another moot point is that it is unlikely to find a single universal formula for equitable distribution of benefits among all the categories of stakeholders, and the differences in livelihood goals, values, resource use and access and other conflicts among various stakeholders should be accepted. It has also to be understood that the involvement of on-site stakeholders is advisable from the very beginning stage of the planning. This is suggested because of the following reasons:

1. This would generate a sense of involvement (which is an essentiality) among the indigenous communities.
2. There is no set formula for equitable distribution of benefits among all the categories of stakeholders because of differences in livelihood goals, resource use and local conflicts.
3. Local factors in general are the most influencing factors in any decision-making process.

What is needed is an approach to ensure and sustain stakeholder participation with opportunity to influence the decision-making at local level under given situations in each case. Often the communities have an inherently weak link in their bargaining power. At times, the corporations may take advantage of the lack of awareness of the commercial value of potential products. Many TK-based commercial products do not fulfil the minimum requirement of patent protection, and the patent system offers opportunities to discount the innovation and contribution of indigenous communities, and the bioprospecting agencies may make small changes in the product to claim novelty.

It is therefore advisable to impart adequate training to the on-site stakeholders in the art of negotiation, and they need to be taught about their strengths and their weaknesses as regards their bargaining power. Concurrently, the process of identifying the potential partners from the list of stakeholders, especially off-site stakeholders, has to go on as they also have decisive roles to play. Broadly speaking, a bioprospecting partnership programme design shall often include the activities such as (a) preparing the biodiversity inventory (including botanical and ethnobotanical collections) of the project area and constant monitoring of its status, (b) capacity building that includes training and infrastructure development in the methods of screening, extraction chemistry and drug discovery and also the sustainable biodiversity management practices, (c) multidisciplinary research on diseases of both local and international significance and (d) equitable intellectual property and benefit sharing arrangements. Besides direct involvement of various on-site stakeholders, these objectives would require involvement of a host of varied organizations and groups for their fruitful completion. Among the government organizations, a broad spectrum could include the forest department, science and technology department, rural development and other government departments and autonomous agencies involved in science and technology, cooperative development and rural development.

Among others, stakeholders with high importance and low power (e.g., local communities, traditional healers, other on-site stakeholders) will obviously become partners. From the off-site stakeholders, the group that consists of the private sector enterprises, NGOs and academics with high importance but high or low power, after evaluating their importance as well as their influences, may be chosen as partners with definite roles to be assigned that should aim to bring the on-site stakeholders from the 'high importance and low power' stratum to the 'high importance and high power' stratum.

The various institutions involved in the partnership work in ways that are mutually beneficial to all parties. It is advisable that the funding agencies are not parties to the research and benefit sharing agreements, and as the government organizations that manage natural resources are generally underfunded, providing direct benefits to government institutions may be a valuable conservation strategy (Rosenthal, 1997a). For example, the benefit sharing plan of the Suriname ICBG project includes contingent benefits to several Surinamese government agencies and the National Herbarium. The benefits accruing from the development of natural products can promote economic opportunities and enhance research capacity in developing countries. As these benefits are shared with the country of origin, it is appropriate to involve the government as one of the partners. Financial arrangements bear more authenticity if the government is involved, especially when the projects are part of grant programmes.

Another significant point is that a successful partnership requires proper communication, honesty and mutual trust among different partners. Citing the case of the Suriname ICBG project, where initiating trust between the parties, particularly between the corporate and private sectors, was a challenge that took almost the full duration of the project, Guerin-McManus et al. (1998) rightly suggest that patience, long-term outlook and foresight of the source country government and the participants are essential in the projects having longer time frame (e.g., those engaged in drug development). Conflicts between different interest groups are a reality in any milieu, and it has to be understood that having a systematic approach to resolve them is foremost rather than let them hamper the prospects of all the partners concerned.

Lastly, the fundamental suggestion of involving women as a group among the on-site stakeholders cannot be overlooked. Another point of great emphasis is that an apt approach in the identification and analysis of the stakeholders and partners in the partnership programme related with the commercial use of biodiversity must be opted as the baseline.

Framework for Partnerships and the Time Frame

Once the groups that have to be involved are decided and the stakeholders are identified, starts the further course of action—the broad framework of the total programme—what in fact it would comprise, what time frame would be needed to reach to a mutually beneficial level of partnerships and who would be the major role players. As was stated earlier, quite complex legal or quasi-legal agreements are being negotiated, for the biodiversity research projects that are often planned on a long-term basis for mutual benefits are taken up with the involvement of several parties. These agreements, as has been indicated in Chapter 6, result from tough negotiations marked with mega-influences attained by the resource-poor and technology-rich side. Therefore, formulating a research programme based on the biological resource of a developing country needs cautious approach and deliberation.

Such kind of deliberations may go on for several months, and maybe years, before a partnership programme is finalized, and therefore, it is foremost that looking to the broader aim, size and duration of the project, an elaborate list of the future activities envisaged is prepared. This exercise will result into a broad framework on which different stakeholders who are enlisted may work in due course to formulate the multilateral agreements. This framework may be in the shape of an 'MOU' between the major role players to show them agreeing to the broader aims and objectives of the project. Deciding a time frame also is possible at this stage for the programme being designed, limiting it to certain specified period with the scope of extension on MAT and the scope of including more agencies at a later stage. Obviously, larger programmes that perceive bioprospecting, drug discovery and research related to endemic diseases as its components may be designed to have a time frame expanding over 10–15 years. The ecological conditions of the particular sites undertaken ought to be the focal point before proceeding to decide the time frame of the programme and the methodology of preparing the inventory of the available biological resources.

Inventory of the Resource

All viewpoints relating to the benefits that could accrue from the biological resources target on a well-laid-out inventory, and it is peremptory that a country or community should have the most relevant and reliable information about the identified genetic resources and their most accurate distribution over the landscape falling within their territory if conservation and sustainable use of biodiversity has to be ensured. On the ground that an inventory of the biological resource would not only provide legal protection from future misappropriation but also facilitate the sharing of benefits accruing from their bioprospecting, it is generally believed that the best way to protect the biological resource of a region against biopiracy is to document it along with the relevant TK (see Chapter 3). It may be reiterated here that any benefit sharing arrangement would critically depend on the ability of the resource provider to link the innovation to its biological origin and to prior knowledge of its uses, and to do this effectively, it is necessary that the ABS regulations enforce the disclosures and submissions in the specification section of the applications (Ghate et al., 1999).

The above outlook is elucidated more when we reflect on the following aspects:

1. If the biological resource that is used commercially is well specified in an inventory along with the parental crop lines or the germplasm accession that is used to breed a new variety or to identify any new chemical entity (active ingredient), and it is also possible to ascertain some cut-off date in the history of the region harbouring the biological resource in natural conditions, the inventory will facilitate in making the ABS regulations more effective.

2. If the inventory is precise to the extent that the most localized source of origin of the resource—for example, a village or village council, an individually owned gene bank or a particular tribe or clan that traditionally cultivate it—is recorded, it would be easier to draft a clear-cut and streamlined benefit sharing arrangement.

3. Whenever the negotiations reach the stage of drafting a bioprospecting partnership agreement, the various features such as MTA, the PIC or MAT for sustained supply of the resource may be effectuated.

To work on the above aspects, it is advisable to create the capability to locate, identify, investigate, collect and compile the inventory about the biological resources, and to safeguard representative samples in situ as well as ex situ in order to utilize this biodiversity in a sustainable manner. The PBRs, though they do not amount to be an alternative to the patent system, may show that the knowledge already exists and thus cannot be patented (Cullet, 1999), enabling patent examiners to conduct more comprehensive *prior art* searches. As a workable option, biodiversity prospecting management institutions may be set up, and the institutions, especially the government undertakings dealing with natural resource management capabilities, can be catapulted and delegated new responsibilities to effectively manage biological resources.

The mere information about a list of the species available would never serve the purpose in a benefit sharing arrangement. What is required is to define the 'ecological bottom line' of the biological resources, and it would be unwise to keep ignoring them. It has been justly advised by Bhojvaid (2003) that the inventory should contain answers to a range of questions as follows:

1. What are the actual ecological impacts of harvesting commercial quantities of MAPs from a habitat?
2. Are some species or resources more resilient to the effects of continual harvesting than others? What can be done to minimize these impacts?
3. What sort of monitoring activities, management practices and silvicultural techniques can be used to ensure that the resources being harvested are not overexploited?

The inventory of the resources that fall under any bioprospecting partnership programme need to answer these questions and encompass all the relevant data in order to ensure sustained access and continued benefits from the resources. Put simply, the inventory should be usable and dependable to formulate an appropriate bioprospecting project in order to ensure the longevity of the venture that is conceptualized.

The process of preparing the inventory of a delineated area should be well planned and executed through deploying appropriate agencies (universities, research institutions, etc.) and training of the personnel involved (taxonomists, community members, traditional healers, students and other researchers). While collection of specimens and preparation of herbarium may be a part of the process of inventorying of the biological resources, as has been learnt from the INBio of Costa Rica's example of facing financial crisis (see Chapter 6), a viable arrangement should be made for the maintenance of the collection of specimens and the herbarium, and any activity that drains out the funds must be curtailed or modified suitably. Brainstorming with the partners involved, an appropriate mode of documentation of the inventory also should be decided before taking on the job.

Formulating the Bioprospecting Partnership Agreements

After having delineated the area to be covered in a project, having clearly defined the groups of people, more specifically the indigenous and local communities who would be its part, and having drawn up the broad framework and the time frame, the identification of various stakeholders and partners should be completed and the biological resource that the partnership programme comprises should be inventoried. When the stakeholders have been carefully identified and the partners to be involved have been chosen, the broad framework and the time frame of the partnership are drawn in the shape of an MOU. This MOU should be the follow-up of bringing together all the items of the expression of interest invited from various partners, who must participate actively from the planning stage onwards. The on-site stakeholders may have to be helped to learn the art of negotiation and to make use of their bargaining power. The exercise of shaping an MOU may run parallel to the making of the inventory of the resource.

Now the stage matures for formulating the eventual contract agreements for partnerships between various stakeholders and partners in the project. These agreements are an important element of the structure for regulating multidisciplinary biodiversity prospecting activities. In fact, carefully drafted contracts between companies and biodiversity collecting institutions have the potential to be extremely useful tools towards ensuring the fair and equitable sharing of the benefits. The agreements also help in building confidence between the visiting researchers and the national or local governments, host scientists and members of the local community (Jermy, 1995). It is well taken that developing first-rate agreements on ABS is a highly technical area, and it requires time and diverse expertise. Often the on-site stakeholders may not appreciate the issues at stake in relation to their own interests. For example, the controversy surrounding two ICBG projects in Mexico and Peru illustrates the sensitivities involved—the concerns of indigenous communities proved very hard to reconcile with the interests of researchers and business (Kate and Wells, 2001). Under the Peru ICBG project, even after the end of the project period, some scientists of the original team continued the research, focusing on potential sources of new therapeutically active agents. This was a precedent-setting project, as the agreements were negotiated with more direct discussions between local organizations, and the resulting scientific network was strong enough to continue it beyond the project period. On the other hand, the Maya ICBG project in Mexico was terminated in the second year of its five years of funding, for it had invited criticisms related to the PIC and an unethical form of bioprospecting. One of the main reasons, as many believe, was that there were no proper representatives on behalf of the indigenous communities to speak out about themselves. (These two cases are elaborated in Chapter 6.) Therefore, building consensus among providers and users of genetic resources and developing trade-offs between their widely differing priorities can be tricky, sometimes proving more beneficial than anticipated, and at times too difficult to take on.

The drafters of agreements (natural resource managers, scientists, community representatives, development practitioners, government representatives, etc.) also generally do not have sufficient commercial and legal experience to negotiate agreements, and as a result, the industrial partners frequently have an advantage in this regard (Rosenthal, 1998). Therefore, proper foundational training and skill improvement programme should be included as a part of the process of formulation of the agreements. This is believed to be an essential component in the formulation of bioprospecting agreements that the holders of TK (along with other on-site stakeholders) should be provided adequate assistance and training in the negotiation, drafting, implementation and enforcement of contracts involving use of their knowledge, as well as evaluating the impacts of those contracts (Tarasofsky, 2002). An appropriate technical training shall help in acquiring legal, administrative and negotiating skills that would forge workable research partnerships. The draft of the agreement should be based on the tenets of the MOU signed in the beginning, and while the general character of benefits can be specified in the MOU, specific agreements can spell out the detailed list of benefits and other operational aspects.

Subsequently, each party may legally scrutinize it to eliminate probable flaws. The basic issues that are generally covered in the agreements include research topics, organisms to be studied, ownership and conditions of material transfer, research investment, patent rights and responsibilities, duration of agreement and the type, amount and recipients of benefits (Rosenthal, 1998). A partnership agreement may include a confidentiality clause also in order to pre-empt disclosure of any information specified in such a clause by either party to any third party for a pre-decided number of years even after the expiry of the agreement. Besides, there may be more number of partners at the time of drafting the various agreements than there were initially at the time of framing the MOU, and this number may not remain fixed—as the need arises, more partners may have to be included and more agreements also may have to be formulated, aiming for innovative and flexible collaborations.

In other words, the project may have multilateral agreements when a multitude of institutions are involved, and if it comprises cooperative agreements, then settlement for grants and other funding may also be required. A few typical agreements, for example, are as follows:

1. Agreement between a university and the community organization for inventorying and documentation;
2. Multilateral agreement between a corporation, a university and the communities for capacity building;
3. Agreement between the corporation or industry, the government and the communities for the bioprospecting programme and institution of the Trust Fund;
4. Agreement between the community organization, the government, local NGOs and other conservation organizations for conservation measures and sustainable management of the resources;
5. Agreement between a funding agency, the government, local conservation organization and the community for the community development works; and
6. Agreement between an individual scientist, researcher or expert and the community organization to meet any other specific purpose or requirement.

Provision may be chosen for advance payments, both monetary and non-monetary, to the resource provider in order to present evidence to the local communities and the government of the commitment on behalf of the partners and also to facilitate early establishment of the 'trust fund'. This may be in the form of the lump sum payments, per sample fees, payment on resupply of samples or in-kind contributions of equipment, training, medicines and so on.

In addition, a reasonable flexibility should be integral to the agreements in order to put in further negotiations at later stages if need arises. Another important element is the timing of royalty negotiations. The best time for a resource provider to work out royalty arrangements may be, as Rosenthal (1997a) suggests, after preliminary data on an extract show therapeutic efficacy and the compound has been chemically analysed. Sometimes, a lump sum payment may act as a bargaining tool, while at times per sample payments may be chosen. Some commercial research partnerships use the tool of MTA, which is narrower in scope and defines the rights and responsibilities related to the specific biological materials transferred. As the resource provider has greater control over the genetic materials offered, this arrangement may be adopted as an alternative to the approach of the transfer of the research results.

The set of agreements should also include the clearly identified and exactly chosen benefit-claimers, and it demands an extra care to ensure that no genuine benefit claimer is excluded from these arrangements. To cite an example, the JNTBGRI did not provide for cash incentives to the scientists in the arrangement of sharing of benefits with Kanis, showing a good gesture. However, it may not be always true that all the scientists involved for all the time to come would be willing to give off something that is genuinely due to them. Instead of who are willing to receive benefits and who are not, the question should be: Who are entitled to draw the benefits, and who are not? Examining these intricacies, the partners of a project may decide to have separate agreements for benefit sharing arrangement and for the commercial R&D schemes.

Two other aspects that are extensively important are related to the terms of partnerships connected to the issue of licensing of an invention and sharing of the commercial gains, and to the PIC of the communities holding knowledge about biodiversity. The collaborators need to converse deftly to delve on these aspects. Wynberg and Chennells (2009) very logically state that it needs to get these aspects right from the start and adopt the fundamental principle of benefit sharing as an absolute one, engaging the communities holding knowledge about biodiversity from the very outset of a project. It is without doubt an utmost requirement that obtaining the PIC with due engagement of the communities as active partners is well ingrained in the bioprospecting partnership agreements.

Elaborating more on this issue, in the context of the San Hoodia case from southern Africa, Wynberg and Chennells (2009) clarify that the negotiating process (that followed an initial crisis showing mistrust of the community to a government agency) between the CSIR and the San demonstrated the importance of relationship building and political climate conducive to fair deliberations, further affirming the prominent role played by NGOs, legal representatives and intermediaries in benefit sharing.

Arrangements for Benefit Sharing in Bioprospecting Projects

In the context of drawing the mechanism for the sharing of benefits that accrue from the commercial use of biological resources to the resource providers in a bioprospecting partnership programme, the types of benefits that would flow to various stakeholders may be discussed in order to facilitate the distribution of benefits equitably. Here the term 'resource providers' is used for on-site stakeholders, more specifically the indigenous and local communities, who are involved in the collection, sustainable management as well as conservation of the biological resources falling within the region included under the project, while the term 'source country partners' is used to specify the resource providers as well as other institutions of the source country wherein the project involves international funding agencies and collaborators.

Types of Benefits to the Resource Providers

There is an array of benefits that accrue to the partners in a bioprospecting partnership. For example, a pharmaceutical corporation may provide equipment for processing samples and scientific training in addition to the future royalties (e.g., in the INBio–Merck collaboration). A research institution may include the indigenous and local communities as co-inventors (e.g., patent applications filed by JNTBGRI with Kanis as co-inventors) that may lead to future earnings from royalty. Glowka (1997) classifies these payments into the following three categories:

1. *Upfront payments*: These are the very first payments received at the beginning of the contract.
2. *Milestone payments*: The rewards received at different stages of research when the partnership is in progress.
3. *Royalty*: A share in the earnings from royalty subsequent to marketing.

While the mode of upfront payment comprises the capacity-building and infrastructure development programmes, providing immediate (short-term) gain, earning royalty from a future pharmaceutical product is a distant (long-term) benefit. The following two kinds of benefits that accrue to the source country partners were identified by Rosenthal (1998) and shall be helpful in working out details of a benefit sharing arrangement:

1. *Advance payments*: Access fees may take the form of lump sum or milestone payments, per sample fees, payment for resupply of samples or in-kind contributions of equipment, training, medicines and so on. Advance payments are also valuable for establishing trust funds that can provide immediate benefits to stakeholders.
2. *Equipment, training and infrastructure*: Commercial partners or nonprofit funding organizations may provide resources to help build the capacity of source country partners to execute current or future needs for bioprospecting research, medical care, biodiversity management and so on.

Gupta (1999b) suggests the following matrix of benefits:

1. Material–individual
2. Material–collective
3. Non-material–individual
4. Non-material–community or collective

The first category of material–individual rewards includes the conventional incentives such as patents, licence fees, contract money and monetary rewards for innovations and conservation efforts. The second category, material and collective incentives, offers enormous scope for experimentation. Several kinds of trust funds, guarantee, risk or ventured capital funds can be set up to promote conservation, value addition, commercialization and so on. These funds should provide enough flexibility for communities to pursue culture-specific norms of conservation as well as offer reward and/or compensation to outstanding local contributors. The third category, non-material individual incentives, includes honour, recognition and respect for such individuals who have contributed extraordinarily to the goals of conservation, value addition or both. The fourth category, non-material collective benefits, includes policy reform, institution building, incorporating local ecological knowledge in the educational curriculum at different levels, development of markets for organic and other local products at national and global level and many more. Many authors regard a patent or other form of IPR as the primary benefit. The other types of benefits as envisaged by the CBD that were discussed in the previous sections need to be brought into an international system.

In actuality, any one type of benefit or a combination of all these can shape the arrangements of benefit sharing under specific agreements.

Capacity Building

Capacity building forms an important component of the upfront payments. It includes technology transfer and specialized training in species identification to enhance parataxonomy and in screening for pharmaceutical uses that may bring about skill improvement among local communities. These may include basic technological skills for extraction, chemical analysis and purification of samples. Other areas can be training in database management and equipment for sample preparation and storage, even for technology of developing natural products and biochemistry research as a part of larger programmes. The capacity-building benefits may also include development of source country infrastructure and institutional alliances for biodiversity management and biomedical research and service delivery

(Rosenthal, 1997a). These may also include provision of vehicles, renovation and improvement of laboratories, community health clinics and herbaria.

Sustainability

In addition to the above, an increased local and national capacity to coordinate conservation and sustainable utilization of biodiversity should also be stipulated in the agreements. The sample collecting bodies mostly include botanical gardens, universities and private corporations that have no direct stake in the resource base. Therefore, a local or national institution should also be involved in order to effect sustainable utilization of the biological resources with a well-laid-out plan for the same.

Another alternative for habitat conservation and sustainable resource management (that is possible with the flexibility of smaller agreements) is to provide subsidies or impose taxes at local level, which could target particular habitats and specified community. For example, a tax at the rate of per kilogram of samples may be paid by the exporters to repopulate the medicinal plants collected by providing seeds and labour or wages. Similarly, the agreement may contain regulations on medicinal plants, specifying the species and parts of the plants that can be exploited and the region of collection and the number of plants that may be collected.

Going further, different instruments and arrangements that have been tested and chronicled in various case studies should be meticulously investigated to manage bioprospecting in the most efficient manner. The NGO community has gained valuable experience over the recent years in implementing policies among indigenous peoples. Quite a few of them have earned recognition among national and international donor agencies and their strength should be taken advantage of while making arrangements for benefit sharing. It is also pertinent to speculate a number of partners if the project is comparatively large or the time frame is longer. In the action guide prepared under the IUCN project on the CBD and the international trade regime, Tarasofsky (2002) very rightly recommends that the parties to the WTO and the CBD should create opportunities for dialogue and contact between TK-holders, the private sector, governments,

NGOs and other stakeholders to enhance cooperation between these actors, at community, national, regional and international levels. As a matter of fact, inter-sectoral coordination being a prerequisite, the three key role players who could facilitate the mechanism and provide expertise are government institutions, NGOs and international organizations.

Looking to the interests of the source country providers, it is also believed that the approach should centre on making local and national scientific institutions more advanced. For example, under the INBio–Merck collaboration, Merck provided training and equipment for the milling and extraction of biological samples, which was conducted by INBio or subcontracted out to laboratories at the University of Costa Rica. On similar lines, in-country extraction and initial screening of the samples may be conducted in the local (national) laboratories (universities, other national R&D institutions and local pharmaceutical companies), taking full advantage of the training and equipment (resulting from an upfront payment). Further screening and product development and marketing may be conducted abroad.

Traditional healers, as the holders of TK, may be enabled and empowered under the agreement to hold patent and joint patent rights with a pharmaceutical partner. For, when ethnobotanical knowledge is utilized in the development of a successful product, a greater percentage of royalties will be accrued than for randomly discovered samples. Similarly, a condition may be put for the exporters and intermediaries to engage sample collectors from the local communities only in order to enable them to enhance their wage earnings.

In addition to the above, many other options are available to actualize the benefits from a bioprospecting programme. A number of additional provisions may be drawn from the following extensive list while working out the arrangements for benefit sharing:

1. A small research agreement may be required if a researcher wishes to collect samples from a forest and seeks information from traditional healers for ethnobotanical research. This may need only three parties: the state forest department, local community organization and the researcher as an individual or his institution on his part.
2. Priority access to any resulting products, and the additional use of samples for in-country screening for treatment of local

tropical diseases, may be envisaged. With respect to the development of TMs for wider local use, certain conditions may be included as an intrinsic part of the capacity building.

3. Market access may be guaranteed for the ultimate natural products in conjunction with a sustained supply of raw material to the industries involved for which market survey may be conducted with a commercially viable partnership. For example, 'The Body Shop' (a cosmetics manufacturer) established 25 trade partnerships (e.g., in Brazil, Ghana, Nicaragua, USA) with local communities in 1997 to meet these aims (Laird and Kate, 1999; Walter et al., 2004).

Lastly, it may be emphasized that there can be no model bioprospecting arrangement, as each project would have a different set of biological resources, each indigenous and local community may have a different cultural tradition and even the local laws may be varying according to its own social goals and priorities. Even the very initiation of a project may be the catalytic effect from different institutions—sometimes by a government institution, at times by an international environmental organization or in some cases, it may primarily be initiated by a small company, and not necessarily by an indigenous and local community or an active NGO. The 'Guidelines on Access to Biological Resources and Associated Knowledge and Benefits Sharing Regulations' notified recently[2] by the NBA of India also provide various options for choosing a set of benefit sharing arrangements.

Other Benefit Sharing Arrangements

There may be times when an agreement needs to be signed in the absence of a full-fledged bioprospecting project for simple extraction of few local genetic resources, wherein there may not be any contribution of the TK. If there is ambiguity as regards the principal beneficiaries, returning benefits to communities in the benefit sharing process can be extraordinarily complicated in practice. In such cases,

---

[2] *Source*: www.nbaindia.org (Accessed in May 2015).

an approach on the following lines may be adopted for limiting the complications involved:

- Local communities who are capable of protecting key habitats need to receive benefits from any exploitation process. The exporters and intermediaries should engage collectors from these communities, thus enabling more gains to accrue to the local communities.
- Researchers may be charged a fee for access to the area where the natural resources may be found. Indigenous and local communities should be able to charge a user fee for anyone wishing to collect flora and fauna in a particular habitat.

Few other aspects that need a serious and elaborate planning in this context may be alluded to at this juncture. One is that a complex and legal or quasi-legal multilateral agreement of long-term nature may have to be negotiated if a number of species have to be extracted commercially from a region for many years, involving product development, ownership issue of IPRs, capacity building and milestone or royalty payments. In smaller projects, the products involved should confine to local markets that are easier to enter and monitor while larger projects may diversify to larger (foreign) markets, which often require heavy capital investment and large product volume and which tend to be vulnerable to product substitution, if such diversification is feasible in terms of sustainable harvests, product quality and investment requirements (Taylor, 1999). It is also advised that the experience and knowledge of funding measures and instruments in support of national biodiversity conservation activities should be shared adequately, and effectual coordination of funding efforts between different stakeholders should be aimed at. A code of conduct may be devised for both the academic researchers and the commercial entrepreneurs for the purpose of implementing the PIC and benefit sharing provisions (Balakrishna, 1998). Also, a percentage of earnings (say royalty) from commercial sales by a licensing partner may be agreed upon in an initial agreement, while a range may be specified, requiring the parties to negotiate different rates over the period of the project on a case-to-case basis.

Regarding the issue of access to the biological resource, it is most appropriate that before removal of any biological resource or use of the TK also associated with the resource drawn from any site, the anticipated utilization should be very clear to both—the country and the custodian of that site. It should be a key principle that the benefit sharing plan starts not after the product is developed but right from the stage of the access agreement (Walter, 2002). It is also suggested that there should be an inter-agency committee (an instrument as stipulated in the Philippines Executive Order 247) on regulating access to biological and genetic resources, to consider, grant, monitor and enforce compliance with research agreements as well as to coordinate further institutional, policy and technology development (Balakrishna, 1998).

Different projects can have varying types of benefit sharing arrangements. For example, one arrangement may facilitate small-scale development projects and fund small grants in a community in which the bioprospectors collect plants and information, while another arrangement may facilitate sharing of benefits with everyone within a country. An illustrative example of facilitating smaller projects and funding small grants in a community is presented by the Shaman Pharmaceuticals (a small US company that uses ethnobotany for pharmaceutical development), who founded the HFC, an independent nonprofit foundation for this purpose (Moran, 2000).

Lastly, it should be understood that though developing good agreements and well-negotiated partnerships engaging numerous agencies are prerequisite for an equitable benefit sharing mechanism, the process is an intricate one. The mechanism for effective participation of indigenous and local communities in decision-making as well as policy planning process is a central idea, and the on-site stakeholders need to be empowered to make decisions regarding method and extent of compensation and the choice of projects. In due course, they must be involved in formulating the mode of flow of the funds and the benefits drawn.

As has been concluded from the issues deliberated in Chapters 3, 5 and 6, strong and suitable regulations for ABS are the most immediate prerequisite in order to facilitate a near perfect kind of partnership programme. Also, as is evident from the case studies described in Chapter 6, a key component of many benefit sharing arrangements has been the establishment of trust funds. In the following section, a trustworthy mechanism of trust fund has been propounded.

Mechanism of Trust Fund

The establishment of a trust fund has been ascertained in most of the partnership programmes as was elaborated in the cases taken up in Chapter 6 and also in many other case studies, in order to channel not only the flow of benefits but simultaneously the flow of incoming fund money too. That is the reason it is stated that a key component of many benefit sharing arrangements involving the commercial uses of biodiversity is a trust fund. In this section, the mechanism of trust fund is detailed out with its basic structure and other important components. The mechanism is chiefly based upon a few major case studies. For example, the FIRD-TM in Nigeria is a textbook type of case that has a well-laid-out constitution for the trust fund so designed that it is replicable in any country. Another fund, the experiences of which can be replicated in other situations after suitable adaptations, is the FPF constituted under the ICBG Suriname bioprospecting project.

The legal mechanism of a trust fund equips channelling of royalties from the collection stage, when it is deposited into the fund, to the actual utilization in local enterprises for community development, managed through an elected board that includes representatives from the community. This mechanism of distribution of the monetary benefits to individuals or communities apparently makes the trust fund model highly acceptable. A flow diagram (Figure 7.2) has been conceived to illustrate the mechanism of a general type of trust fund for suggesting the control and channelling of benefits from the commercial uses of genetic resources. The basic framework for the establishment of a trust fund is discussed in the following paragraphs.

Constitution of a Trust Fund

On the lines of several trust funds in existence today as an intrinsic part of benefit sharing arrangements under the bioprospecting partnerships, it should be advisable to have a written constitution for the trust fund. A trust fund should generally involve the local people, the government, technical experts from scientific institutions and representatives of the major partners in the bioprospecting projects. If it is

Figure 7.2:
Mechanism for the trust fund

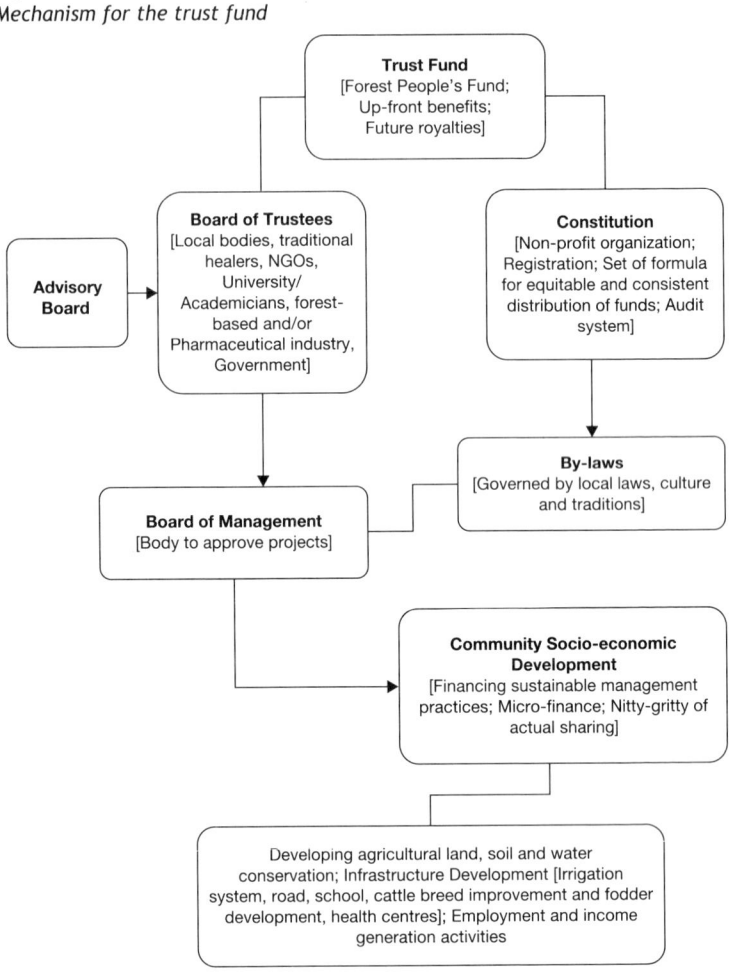

large-size fund, the 'best practice' that may be suggested is to have the following three principal organs:

1. Board of Trustees
2. Advisory Board
3. Board of Management

These three organs form the basic structure of the trust fund in the most typical example of a trust fund mechanism in the case of FIRD-TM. Each of these units has well-defined roles and functions. The top-level body of the board of trustees is to bear the overall responsibility of the affairs of the fund. It may provide patronage to the schemes of things and intervene whenever an issue of conflict arises. The advisory board plays only an advisory role to the fund managers through guidelines and executive orders, if required, in line with the constitution of the trust fund.

The board of management is generally an elected board that includes the representatives from the community. With the responsibility of everyday affairs of the trust fund, its responsibilities include expenditure accounting, fund inflow and budgeting, review of the proposals for socio-economic and infrastructure development projects, biodiversity conservation schemes, usual monitoring and evaluation of projects and so forth.

The San Hoodia case from South Africa is another example of benefit sharing involving a trust fund mechanism. The San community instituted the 'San Hoodia Benefit Sharing Trust' that composed of nominees from South Africa's CSIR and the Department of Science and Technology, the South African San Council (San Council) and from WIMSA. The trust aims at knowledge sharing and also using the fund for general development and training.

The inflow into the fund has provision of milestone payments, and it also includes the share in royalty. The plan of the fund stipulates to tap into only the interest generated and invest it in education and other development projects for which land would be bought and clinics and other infrastructures will be brought up. It is also noteworthy that a simultaneous endeavour was going on for enacting a law for the conservation and sustainable use of biodiversity, and the Biodiversity Act was enacted finally in the year 2004 by the South African government. The Act has provisions for applying for bioprospecting permits, and this enactment has a contextual similarity with India's Biological Diversity Act of 2002.

Another feature that may be an important constituent in the trust fund mechanism is that some of these funds may operate at the regional level, while others may be implemented at the community level. Often the more focused small-scale development projects designed to suit local needs and reach the appropriate populations confined to a few villages may be of more utility, though there can be cases where the system of 'global royalty' that envisages benefit distribution even for all the population of a country may be advisable.

While the local priorities and the location-specific socio-economic situations would be the guiding factor for preparing the framework for constituting a trust fund, the basic framework for the establishment of any typical trust fund may be outlined as follows:

- It is advisable to have a written constitution for the trust fund.
- A trust fund may operate at the community level with focused small-scale development projects confined to a few villages.
- It may also be implemented at the regional level if the scheme envisages the distribution of a 'global royalty' for everyone of a country.
- The socio-economic and infrastructure development work to be done out of the fund money should include efforts for community development, enhancement of employment, general improvement in quality of life, renovation of laboratories, community health clinics and so on, with the aim of sustainability for the programme.
- Health-care research facilities also may be provided to remedy local epidemics and other diseases.
- While about 60 per cent of the fund money may be kept in fixed deposit for earning interest only, the remaining 40 per cent may be used in a specified proportion for the benefit of the on-site stakeholders involved.

Socio-economic and Infrastructure Development Work

The socio-economic and infrastructure development works undertaken from the fund money have their bearing over the continuity of the bioprospecting project. Basically, it includes efforts for community development, enhancement of employment and general improvement in quality of life, and it may include provision of vehicles, renovation and improvement of laboratories, community health clinics and herbaria and so on with the overall aim of sustainability for the programme.

As part of its socio-economic development programme, health-care research facilities also may be provided to remedy local epidemics and other diseases for the benefit of the indigenous and local community. Anything that meets a local need may be taken up. For example, the Kayapo Indians of Brazil extract Brazil nut oil (having hair conditioning properties) locally for the Body Shop, a European cosmetics manufacturer, and there is a second-hand Cessna airplane provided by the company for emergency flights to hospital, a unique step for a common good.

About 50–60 per cent of the fund may be kept in fixed deposit, of which only interest shall be used. The balance of 40–50 per cent may be used in fixed proportions on different types of the work, taken up as per a well-laid-out programme after due approval, and the various proportions of this expenditure fund could be something like:

- 20 per cent for the biodiversity conservation;
- 10 per cent for educational purposes;
- 30 per cent for the traditional healers' associations for group projects or micro-credit funding;
- 30 per cent for community development associations for village projects;
- 5 per cent for women, especially widows; and
- 5 per cent for children's welfare.

One would quite often come across several unusual and unique examples of benefit sharing. One such example is from the Central American country, Belize. After working on the Belize Ethnobotany Project (1987–96), a book was published: *Rainforest Remedies: 100 Healing Herbs of Belize* (Arvigo and Balick, 1998). This book was the result of nine years of interaction of the authors with the traditional healers of Belize, and in an interview (Balick, 2006), author Michael Balick spoke that they (the authors) dedicated the royalties back to the traditional healers, and since its publication, the book could provide $26,000 to the 11 traditional healers who shared their knowledge with the authors, and with this help, the healers could begin their family-level economic development efforts, raised medical plants and some of them started teaching their grandchildren about the uses of plants. Interestingly, it is learnt that the publisher, Lotus Press, also contributed a part from the proceeds of this book to the Traditional

Healers Foundation of Belize, and this book also became a reference for ecotourism industry there. A pension programme also was devised for the 11 traditional healers who had shared their knowledge (Balick, 2007).

The above is only an illustration, and it may vary depending upon the local priorities and the location-specific socio-economic situations.

To sum up, for forging any partnership programme, it is imperative to link the on-site stakeholders closely with the ecological region of the biological resources that are to be used commercially, to delineate the project area and also to have the resources inventoried in a properly defined system. It is also crucial to identify and involve all the stakeholders and the government from the very beginning for long-lasting success of the programme. To begin with, engaging all the stakeholders, an MOU should be prepared, and then should follow the drafting of a set of agreements between different partners of the programme. Concurrently, for channelling the flow of funds as well as the benefits accruing in the project, a trust fund should be established that should follow a written constitution based on local laws, customs and traditions. The drafters of the multilateral agreements may innovate to include novel ways of benefit sharing, and a code of conduct also may be prepared at a later stage to coordinate technology development in future.

8

Bargaining Power of the Resource Provider Countries

The standing of the biodiversity-rich countries (also termed as the resource provider countries and often, in short, 'source countries') becomes weak in the context of the Agreement on TRIPS, for it does not bar a person to claim patent rights in one country over an innovation based on the genetic resources (and associated TK) acquired from another country. The provisions under the Agreement on TRIPS do not support the obligations under the CBD, and as a consequence, the source countries literally have to face the adversity of biopiracy.

The discussion here on the bargaining power of the source countries is aimed at overcoming this adversity and the inequality these countries face today under the current international IPR regime. If the trend of depriving the indigenous and local communities from an equitable sharing of benefits drawn from the commercial use of biodiversity continues, it would undermine the efforts to conserve biodiversity. Also, it would not suffice only to provide for an equitable sharing of benefits from the commercial use of the components of biodiversity, as it would be an unrealistic assumption that the sharing of benefits will ensure a sustainable exploitation of the very resource on which they depend for their livelihood and for meeting their other day-to-day needs. If the resource is in high demand, the exploitation may not remain sustainable and therefore it would be wrong to assume that they would have a greater incentive to help ensure conservation once an equitable sharing of the benefits is arranged. Here comes the role of the mechanism of benefit sharing, and this mechanism will be more robust if the benefits drawn are enough to sustain a particular habitat of the biodiversity. Quite naturally, in order to draw adequate benefits that are proportionate to the value of the resource, the source countries have got to have a relook on how to bargain maximum from the resource-user countries.

Another point of great concern here is that the source countries still have to face the unsupportive international IPR system, for even after more than a decade of deliberations at the global level, inventions derived from genetic resources are being patented by third parties, raising questions as to the relationship between the present IPR system and the conservation and sustainable use of biodiversity and the equitable sharing of benefits. At this stage, it would be pertinent to make an assessment of their bargaining power in the context of negotiations for bioprospecting partnerships to be entered into with the pharmaceutical and other industries of the resource-user technology-rich countries, termed in short as 'user countries'. In order to go ahead with this assessment, it would be relevant to enlist the specific reasons causing inherent weaknesses in the bargaining power of source countries. Some of the significant ones are listed as follows:

1. Many TK-based commercial products do not fulfil the requirement of novelty because they are not new to the communities that supplied the knowledge about them. The companies often simply extract the chemical of interest, and patent offices often conveniently ignore this. Even in the areas where a particular species is endemic to, the TK is not acknowledged and the indigenous and local communities remain unaware of the worth they are deprived of.

2. A scenario where the extraction of a particular genetic resource from a well-defined geographical location, very much drawing upon the TK associated with the resource, gives the lead for drug discovery, there is no foolproof mechanism in the present IPR regime that can ensure an ethical way of recognizing the rights of the local community or an equitable sharing of the benefits drawn. The patent system offers opportunities to discount the innovation and contribution of indigenous communities, and the bioprospecting agencies may make small changes in the

product to claim novelty. The scenario may favour a bioprospecting agency as the indigenous and local communities may not be equipped enough to monitor or survey the patent regime even though the species are originating from their habitat.

3. Many plants of therapeutic interest grow, or could grow, in a number of different countries or districts, meaning that the resource users can take advantage of the lack of awareness of the commercial value of potential products among some communities and get bargain base prices. The situation may worsen if the country's ABS regulations are complex, which may inhibit a user country to move towards a country (or district) where these regulations have more favourable setup. Also, if the regions where endemic species occur are not so well defined, the indigenous and local community may find difficulty in claiming benefits.

4. Patenting is a costly business, and if the invention is appealing enough to be challenged, the real cost in putting a challenge could be monstrous. Moreover, if the loser of a patent claim is a small community, it may not have adequate resource to defend its claim.

5. Another reality is that the holders of TK are finding no takers in the younger generation, leading to a situation where the TK that evolved through several generations is not likely to be transmitted to the newer generation in the manner it has travelled so far. As a consequence, there is no scope even for further additions to the wealth of TK. The approach to directly benefit the holders of TK also is not accepted unanimously in many collaborative programmes.

6. In the absence of a proper IPR regime that could match the flow of the genetic resources and the benefits drawn in the international market from these resources, the source countries are dependent merely upon the existing mechanisms for negotiating, implementing and enforcing bioprospecting contracts. The source countries have been lacking behind in the professional experience in applying or developing workable legal framework in the fields of patents and contracts. However, during the last 10–15 years, many countries of the tropical regions have initiated the process of developing appropriate legal framework for sui generis system for legal protection of their rights and for regulating the ABS arrangements.

7. A major issue for many countries is the lack of market linkages and the information about the value of genetic resources, including the nature and extent of trade in and movement of genetic resources and associated TK (ICIMOD, 2010), leading to the situation wherein the user countries hold an upper hand to determine the economic value of the resource in use, often comparing it with other similar products patented under the international IPR regime.

8. At the negotiation table, the gap in technological capacity and the specialized skills as well as the literacy levels of the indigenous and local communities pose great challenges to be faced by source countries.

9. Proper monitoring of the collection and use of the genetic resources, complying aptly with the agreements reached upon, also is an elaborate task that requires proper costing. Without proper monitoring, the agreements tend to have aberrations in the sharing of benefits.

It would be appropriate to follow this analysis of the weaknesses of the source countries with a brief review of the limitations in the existing international conventions vis-à-vis the strengths and weaknesses of the source countries. In the context of creating a sound and suitable international legal setting wherein the source countries are enabled to protect the genetic resources and the TK associated, it is witnessed almost all around the world that newer developments are happening. New legal instruments are emerging, deliberations are going on, ideas about documentation of the TK are being implemented and, simultaneously, bioprospecting contractual agreements also are being signed and carried forward. These new developments are discussed, supported with few examples, in the following section.

Protection of Traditional Knowledge and Genetic Resources of the Source Countries

The weaknesses of the source countries in the context of international IPR regime are explicit from the above discussion, and concurrently we realize their strengths also in so far as the unexplored wealth of biodiversity is concerned. If this realization works, and the various debates at international level reach the finishing line, it would be possible to evolve an IP strategy that does not leave any policy or legal vacuum that is detrimental to the commercial and sustainable use of biological resources of the tropics.

It is well perceived by several governments, organizations and biodiversity activists that to avoid the conflict between the technology-rich 'North' and the biodiversity-rich 'South', an amendment of the Agreement on TRIPS to accommodate some essential elements of the CBD is crucial. India has proposed, in this context, that under the Agreement on TRIPS, patent applicants should be required to disclose the source of origin of the biological material utilized in their invention and should also be required to obtain PIC of the country of origin (WIPO, 2003a). If this is done, it would enable the domestic institutional mechanism to ensure sharing of benefits of such commercial utilization by the patent holders with the indigenous communities who have the claim over the TK that has been used. In order to clarify the status of TK, India's Patents Act 1970 has been aptly amended. Section 10 (Contents of specifications) of the Patents Act of India, 1970,[1] as amended by the Patents (Second Amendment) Act 2002, provides that if the applicant mentions a biological material in the specification, the application shall be completed by depositing the material to an international depository authority under the Budapest Treaty and by fulfilling the condition of disclosing the source and geographical origin of the biological material in the specification, when used in an invention. Also Section 25 (Opposition to the patent), as amended, allows for opposition to be filed on the ground that (a) the complete specification does not disclose or wrongly mentions the source or geographical origin of biological material used for the invention, and that (b) the invention so far as claimed in any claim of the complete specification is anticipated having regard to the knowledge, oral or otherwise, available within any local or indigenous community in India or elsewhere.

While several governments, organizations and biodiversity activists believe that an amendment of the Agreement on TRIPS is

---

[1] *Source*: www.ipindia.nic.in/ipr/patent/patent_Act_1970_28012013_book (Accessed in June 2014).

imperative to accommodate some essential elements of the CBD, as discussed above, no biodiversity-rich country is waiting for the same. Effectually, many initiatives have been undertaken the world over to document TK with a kind of self-assurance that the range of well-drawn inventories and databases, collections and PBRs and other forms of documentation of TK would help in forbidding its misappropriation and assist in claiming benefit wherever there is an indubitable use of the TK held by a community. And yet there is an apprehension that the documentation of TK in general, if made readily available, could lead to misappropriation through some hidden ways and means.

This is the reason why it is expected that countries such as India, being one of the major exporters of NWFPs, especially MAPs, should exercise her bargaining power in coalition with other major suppliers of these products in the WTO renegotiations. Sharma et al. (2001) indicate that for commercialization of the biological resources, robust market demand, adequate product availability and appropriate prices generally provide the strongest incentives for harvesters, buyers and processors. Nevertheless, the protection under the international IPR regime is the need of the day, or else this robustness may not match the same as witnessed in the trade practices of large pharmaceutical corporations, especially when a phenomenal growth of the sizes of these corporations is emerging quite expressively.

What is argued further is that because of the ongoing misappropriation of TK across the world, in any case, there is a strong need to establish an effectual set of rules at the international level. There is a general agreement among the WIPO member states also to develop an international legal instrument that would give TK adequate effective protection. An IGC was formed in 2001 for the purpose, though WIPO started to work on TK in the year 1998, listening directly to TK-holders, learning of the needs and expectations of some 3,000 representatives of TK-holding communities in 60 locations all over the world (WIPO, 2004a). However, although mainly developing countries made the first move and pushed the negotiations, the discussions are not neatly divided along 'North–South' lines, and communities and governments do not necessarily share the same views (WIPO, 2012b). And yet the WIPO covers different programmes through workshops, training and many other modes on the technical aspects of IP management. The IGC has comprehensively been working on the protection of TK through existing IPR regime and besides that on the development and application of sui generis systems.

The scheme of things under the auspices of WIPO includes facilitating a normative process among member states aimed at developing an international legal instrument, providing complementary capacity strengthening, and maintaining inter-agency and external relations with other entities or treaties such as WTO, UNESCO, CBD, FAO and UNCTAD. The capacity-building activities of WIPO aim at making an effective use of the IP systems and are extended to governments and NGOs, indigenous and local communities, research institutions and so on, and the offers include practical assistance as well as technical advice. However, the tropical country members have many more expectations from the WIPO, especially the IGC, and some day, they may get what they expect once the developed 'North' realizes what is best suited for the overall conservation of the world's biodiversity.

Prior Informed Consent: A Policy Vacuum

While the negotiations are on the run, the issue of obtaining consent of the indigenous and local communities before initiating the commercial use of the genetic resources that they are custodian of is a matter of concern. It is presumed that the TK-holders must be consulted before their knowledge is accessed or used by a third party and an agreement should be reached on appropriate terms. Yet the legal instruments that exist today are not compatible with international scenario, and a policy vacuum is very much obvious.

As a result of a discussion of the WIPO bodies on these lines, the IGC was set up to go into the relationship between IP, Genetic Resources, TK and Folklore. In one of the initial meetings (May 2001), the USA made clear that while the IGC could consider and discuss benefit sharing, it would not agree to the discussion of the concept of PIC in terms of access to genetic resources and IPRs (Raghavan, 2001). This presents the fear of biopiracy as a reality. The USA, not a signatory to the CBD but only an observer, has used proxy to hold up and weaken the bio-safety protocol, though Europe and the Group of 77 claimed it as a major achievement. Attempts in the Council on TRIPS to focus on these questions, and particularly to require that those seeking to protect IP based on genetic resources should be asked to identify the source, and also that the patenting authorities in any country should make this a condition before entertaining requests for IPRs, have been strenuously opposed by the USA. The UNEP and the CBD have not been able to deal in depth with these issues, and attempts in UNCTAD to discuss these in open forums have been blocked by the USA (Raghavan, 2001).

This is not the only vacuum in the benefit sharing arrangements. For example, most of the bioprospecting partnership agreements began during the last decade of the twentieth century, and it is noteworthy that in the follow-up of Article 15.7 (*Access to Genetic Resources*) of the CBD, none of the countries from where the case studies were picked up for review had enacted any distinct ABS laws or regulations. Article 15.7 specifically stipulates that each CP shall take legislative, administrative or policy measures with the aim of sharing in a fair and equitable way the results of R&D and the benefits arising from the commercial and other utilization of genetic resources with the CP providing such resources. Such sharing shall be upon MAT.

What is depicted from the various benefit sharing experiences is that as a result of non-enactment of suitable ABS laws and regulations, every case study was an experiment of its own kind and was seen with apprehension. The BIO, a global not-for-profit trade association representing about 1,100 companies, universities, research institutions, investors and other entities dealing with biotechnology in more than 32 countries, very logically states (BIO, 2013) that 'only legislation that takes into account the interests of the potential users, as well as the providers of genetic resources, in a balanced way will stimulate the environmentally sound use of biodiversity and secure adequate benefits from their exploitation'. So long as such legislation was not in existence, the apprehension persisted and the benefit sharing experiments were targets for severe criticism. As in one case of Kani tribes in India, the royalty of 2 per cent was criticized as being too small with regard to the contribution of the Kani, though it was reported that this was the amount that was prescribed in the standard format for transfer of technology agreements prescribed by the CSIR of India. It is a recent phenomenon that several tropical countries now have or are in the process of making relevant laws for ABS.

Incidentally, a very disappointing phenomenon was reported by Dalton (2004) that shows low level of interest of large companies in bioprospecting. Dalton reports that some companies such as Monsanto and B-MS have shut down their natural products divisions entirely, and another manufacturer, Merck & Co., has now halted investment in the Costa Rica project, providing its last grant of $130,000 in 2001. Though officials of these firms refused to tell why they have ended their bioprospecting programmes, bioprospecting advocates believe it is because of the lack of a firm framework for benefit sharing between host nations, scientists and commercial companies.

The ambiguities in the international conventions, guiding the benefit sharing principles, are ought to be removed. Also, a more valid reason for low level of interest of large companies in bioprospecting could probably be purely economic, and not procedural. The nondisclosure of the reason on the part of these firms could be attributed to the attitudinal protectionism that is common in manufacturing industries. The scenario reflects more of the conflict between the developed and the developing countries. This issue has been alluded to in Chapter 1, and the hard fact is that even after about 20 years of the existence of the Agreement on TRIPS, the debate is still on to synchronize its provisions according to the tenets of the CBD.

The issue of PIC is elaborated especially with some handy tips and practical aspects in Chapter 9.

Need for Tough Negotiations

When the international IPR regime is considered with respect to the interests of biodiversity-rich countries, it is found that the most encouraging instrument in their hands is the CBD. The oft-quoted objectives of the CBD (conservation of biological diversity, sustainable use of its components and the fair and equitable sharing of the benefits arising out of the utilization of genetic resources) also include equitable sharing of the benefits by appropriate access to genetic resources, appropriate transfer of relevant technologies and appropriate funding. But the provisions or the lack of provisions in the Agreement on TRIPS with regard to the interests of the biodiversity-rich 'South' are so much polarized in favour of the technology-rich 'North' that negotiations at the international level are still too uncompromising and intricate to come to an end.

Particularly, in the perspective of the 'technology transfer versus resource transfer', it may be noted that in the early 1980s, the top 20 pharmaceutical companies held about 5 per cent of the world prescription drug market, and two decades later, the top 10 companies control 40 per cent of the market (Mehta, 2002). Also that 500 corporations control 70 per cent of world trade, 80 per cent of foreign investment and 30 per cent of world GDP, and the transnational corporations, the powerful business enterprises controlled by centralized hierarchies, pursue their own interests and owe loyalty to no community, government or people (Miller, 1995). It is a well-known secret that these corporations are able to influence or modify the policies of large industrialized states. This indicates that the resource-poor and technology-rich side is becoming more powerful economically and is, as a result, attaining mega-influences.

There is another point. The USA was the only major country not to sign the CBD at Rio de Janeiro, and President Bush claimed that accession to the CBD was impossible because the USA could not make commitments for its private industry to transfer protected technology (Miller, 1995). This only corroborates the mega-influences of the technology-rich side. Opposite to this, the technology-poor side is moving, although slowly, from non-influence to near-influence. For example, India's share in global trade is slightly over half a per cent, yet it has emerged as an influential voice in the international trading community (Mehta, 2002), and this voice should be able to provide leadership to the fellow developing countries. These countries may try a coalition and persuasion strategy to pursue their interests. In the quid pro quo of this nature across the board inevitably, India may have to push its own agenda, taking advantage of the provisions of the CBD related to the access to genetic resources, TK and benefit sharing objectives, thumping on the issue of transfer of technology and appropriate funding.

But what has been observed conversely during this period is that the countries of the 'South' are equivocally more involved in formulating strict ABS regulations, Biodiversity Laws, and other sui generis systems advocating protectionism on their part. This situation leads not only to more and more tough negotiations at the international level in the context of conservation and sustainable use of biodiversity, but

also to a very uncalled-for adversity of hampering research in the field
of many endemic diseases in the poorer countries of the world. The
'*Preamble*' of the CBD speaks of the conservation of biological diversity
as a common concern of humankind and the importance of biological
diversity for evolution and for maintaining life-sustaining systems of
the biosphere. But proper understanding of these tenets is still too far,
and it is still unclear when there shall be a consensus on these issues.

As regards the negotiating power of the indigenous and local
communities, particularly when pursuing bioprospecting partnership programmes, it is quite obvious that one has to leave apart the
conflicts discussed in the previous sections and forge ahead, irrespective of the global patent regime, on the lines the various case studies
depict having chosen in order to draw the benefits from sharing of the
genetic resources and the TK associated with these resources. While
negotiating at global level, on the part of the source countries, a set
of the following steps might help in enhancing the strength of the
indigenous and local communities:

1. The local laws should provide for recognition of the rights of
   indigenous and local communities on TK associated with the
   resource;
2. There are adequate legal instruments to specify the source of
   various TK components;
3. The capacity building of indigenous people is supported in
   order to negotiate with interested parties in so much as to enable
   them to carve out the further set of negotiations in future;
4. The role of indigenous and local communities should be recognized in so far as conservation and sustainable use of the
   genetic resources are concerned; and
5. The indigenous communities are assisted by the government in
   the negotiation process in order to develop realistic agreements.
   (As has been mentioned elsewhere, the role of the government
   is more than that—the involvement of the government as one
   of the partners in any or more than one partnership, especially
   the constituents that involve any kind of finances, add to the
   credibility of the programmes.)

The various options suggest that we need community rights to protect and promote the local management of biodiversity, and shield

local innovation from the encroachment of the industrial IPR system. As regards the strength that the CBD is able to provide to the developing countries in forging negotiations for technology transfer and equitable benefit sharing, it is accepted that the CBD undoubtedly is a well-intended document. Articles 15–19 (access to genetic resources, access to and transfer of technology, exchange of information, technical and scientific cooperation and handling of biotechnology and distribution of its benefits) address exchange and cooperation with regard to science and technology. Access to the genetic resources on MAT, scientific research based on genetic resources with the full participation of CPs, access to and transfer of technology on concessional and preferential terms where mutually agreed, effective participation in biotechnological research and taking into account the special needs of developing countries—all these provisions support this view. However, this language is very general, particularly when IPRs are involved; access and transfer should have corresponded strongly with these rights and should better have been consistent with well-defined terms and conditions. Article 16.2 of the CBD that deals with access to and transfer of technology states:

> Access to and transfer of technology… to developing countries shall be provided… under fair and most favourable terms, including on concessional and preferential terms *where mutually agreed.…* In the case of technology subject to patents and other intellectual property rights, such access and transfer shall be provided on terms which recognise and are *consistent with the adequate and effective protection of intellectual property rights.*

As the parts in italics show, the access to and transfer of technology is a subject that shall be subservient to the IPRs, more specifically if the technology is a patented one, it may be transferred if and only if the CP agrees to it. Similarly, Article 16.5 reads as follows:

> The Contracting Parties, recognising that patents and other intellectual property rights may have an influence on the implementation of this Convention, shall cooperate in this regard subject to national legislation and international law *in order to ensure that such rights are supportive of and do not run counter to its objectives.*

Patents are awarded under written laws, and unless some title is provided to a party for a share in the benefit, it cannot be shared. And if it does not authorize sharing in well-defined legal terms, at least a patented and protected technology cannot be transferred. Therefore, the US President is right when he claims that he could not make commitments for private industry to transfer a patent-protected technology. Nevertheless, if other developed countries became signatory to the CBD, it might have been that they footed their stand on the weakness of the language addressing technology transfer. Miller (1995) very rightly mentions that the language in these articles (Articles 15–19) tries to do the difficult task of satisfying the interests of both the developing and the developed countries; the language that calls for transfer of technology was intended to satisfy the interests of the developing countries, and the qualifiers that deal with mutual agreement and IPR were included to satisfy the developed countries.

Finally, we are left with a bare ambiguity, and the technology transfer remains an issue to be constantly pursued by the developing countries. An analogy is often drawn between the stronghold of oil producing countries over the world economy and the probable option values that the tropical countries have because of their biological wealth, particularly in the forthcoming era of biotechnology. It all depends on how strongly the resource-rich and technology-poor 'South' is able to negotiate crossways with the resource-poor and technology-rich 'North'.

Is Intellectual Property Not for Small People?

There is an apprehension that the advocates of the indigenous people are pleading for empowerment that is unachievable. Nevertheless, on close examination, we find that there are a few factors, as shown below, that may prevent the exposure to the IPR regime, and simultaneously, there are alternatives to remedy the situation:

1. The factor of illiteracy: Not impossible to overcome.
2. The limitation of language: Finances should be managed for making available local language databases on patents and the TK.
3. Patents an expensive affair: The objectives of having patents have to be reinvestigated, ethics has to be included in the economics of IPR regime and laws and regulations have to be modified.

Summing up, in the context of an equitable sharing of the benefits accruing from the biological resources and the associated TK, it may be stated that the bargaining power of the resource provider countries of the 'South' revolves around aiming at the process of overcoming the discrepancies in the current IPR regime. In the light of ever-growing demand for natural products the world over, the benefits drawn from the biological resources have to be used to sustain a particular habitat and the source countries may have a relook on how to bargain from the resource-user manufacturing sector. It is frequently read as suggestions in literature that there should be more of debate and discussions and studies on these issues. With more awareness among the mainstream of polity in developing countries and more effectiveness among the NGOs involved in resource management, we can expect happening of INBio–Merck type of negotiations in each community. The incentives in these programmes may focus on the empowerment of local communities so that they may enhance their negotiating skills. Also, we may need 'barefoot ABS practitioners' who (a) are educated in the IPR laws, (b) have better knowledge for conservation and better understanding of the future prospecting of local resources, (c) have studied cases on benefit sharing and (d) are skilled in the art of negotiations. Obviously, the developing countries and the indigenous people who live on the fringes of the forests in those countries sustaining on MAPs have enormous bargaining power that they are not being allowed to use. But this situation may not be going to last long.

9

Resolving the Issues: The Way Ahead

Looking at the direct linkage between biological diversity, human health and livelihood options of forest dwellers, the central idea behind the theme of this book has been to find out what could be best suited to all stakeholders involved in the commercial use of biodiversity. The theme is in focus in almost all the chapters, more specifically in Chapter 6 in which few of the most significant case studies related to commercial uses of biodiversity have been reviewed. The prime intention in this chapter is that the finality should turn out to be a functional tool for the indigenous and local communities, foresters, government agencies, funding agencies, NGOs, research institutions and even for the pharmaceutical industry if they enter into the trade involving commercial use of the components of biodiversity.

While trying to work out the functional tool referred to above, it is found that the global scenario is still abound with several stumbling blocks and not many examples of partnership programmes are found to transpire today. This is despite numerous treaties and conventions that exist today such as the CBD, the Cartagena Protocol on Biosafety to the CBD (see Box 9.1), the Bonn Guidelines on Access to Genetic Resources and Fair and Equitable Sharing of the Benefits Arising out of Their Utilization (see Box 9.2) and the Nagoya Protocol on Access to Genetic Resources and the Fair and Equitable Sharing of Benefits Arising from Their Utilization to the CBD (see Box 9.4), and many other national laws or regulations related to the ABS issues. This aspect has been discussed candidly and scrupulously in the previous chapter too. Keeping in view the outcome of

Box 9.1:
Cartagena Protocol on Biosafety

The **Cartagena Protocol on Biosafety**, a supplementary agreement to the CBD, is an international treaty adopted in Cartagena (Colombia), which focuses specifically on trans-boundary movement of any living modified organism resulting from modern biotechnology that may have adverse effect on the conservation and sustainable use of biological diversity of a country. The Protocol ensures that countries are provided with the information necessary to make informed decisions before agreeing to the import of such organisms into their territory. Adopted and opened for signature in 2000, the Protocol became effective in September 2003. The COP to the CBD serves as the Meeting of the Parties to the Protocol (COP–MOP) and acts as its governing body.

Source: http://bch.cbd.int/protocol/background (Accessed in September 2014).

Box 9.2:
Bonn Guidelines

The **Bonn Guidelines on Access to Genetic Resources and Fair and Equitable Sharing of the Benefits Arising out of Their Utilization** are the most important feature within the arena of the CBD. These guidelines provide the best expansion to the scope of the CBD. The ABS regulations and the biodiversity laws of many tropical countries seem to have grown parallel to the progress of these guidelines, which are so comprehensive that these can be fitted aptly within the biodiversity laws and ABS regulations of any country of the world.

The elements of these guidelines, though not legally binding, prepare the CPs to the CBD with clear and certain authority of an all-inclusive will to tackle the important issues that involve (SCBD, 2002):

- ABS process;
- PIC of providers;

- Identifying basic requirements for MAT;
- The main roles and responsibilities of users and providers and other stakeholders;
- The incentives, accountability, means for verification and dispute settlement aspects;
- The elements for inclusion in MTAs;
- The types of benefits—both monetary and non-monetary; and
- Meet all the requirements for the common good.

The guidelines are quite elaborate on the above aspects, and if the existing regional and local legislations and regulations as well as the deliberations of the WIPO on issues relevant to ABS are taken into account, the set of necessary legal instruments is replete with what is needed for the protection of TK and for an equitable commercial use of the genetic resources of the tropical countries.

A competent national authority equipped with necessary financial resources also is envisaged under the guidelines that should act as a national focal point for ABS in order to promote awareness on implementation of relevant provisions of the CBD and facilitate effective transfer of appropriate technology to the resource-provider stakeholders and indigenous and local communities of the developing countries, in particular the least developed countries and the small island developing States. Although the adoption of the Nagoya Protocol introduces the concept of benefit sharing in a much stronger legal perspective at the international level, the CPs to the CBD have asserted that the ABS provisions of the CBD and the Bonn Guidelines remain pillars of international ABS regime (Morgera et al., 2014).

the account of case studies in Chapter 6, the discussion now may be summed up to find the best possible ways to draw maximum benefits to all the stakeholders.

Before making suitable recommendations for the best possible ways, we may look for the probable reasons as to why not many examples of partnership programmes are found to transpire today. The fundamental reason could be that embarking upon any partnership programme is a really tough task. It involves many off-site stakeholders having diverse financial interests and many on-site stakeholders forming diverse groups and having different tenurial rights and conflicts of interest. It also involves academic and research institutions that are not comfortable with intricate kind of negotiations that would be needed in any bioprospecting partnership project. Another observation, as in India, is that in many states, the programmes launched for the benefit of indigenous and local communities, in which the collection or trade of NWFPs and MAPs is a part, are aimed basically to draw benefits as wage earnings only and not as other upfront or milestone payments that could accrue to these communities if bioprospecting was chosen as the mode of trade and development. Institutions such as the Madhya Pradesh State Minor Forest Produce (Trade and Development) Cooperative Federation (this institution has launched a well-known brand of 'Vindhya Herbals' for its herbal products[1]) and 'Oushadhi' (a Government of Kerala Undertaking[2]) are in the trade of herbal products, and though value addition to the products, standardization, branding and good marketing interventions are steps that ensure benefits to the forest dwellers engaged in the collection of MAPs, the benefits do not measure up to what could be achieved if these interventions scaled up to bioprospecting. The reason for not doing so yet again is that the task is tough.

Before deliberating further on this issue, it would be relevant to recapitulate the main findings that have been brought up in this book. In a nutshell, the salient features of the trade of the biological resources may be enlisted as follows:

1. The biological or genetic resources are crucial to the forest-dwelling indigenous and local communities in developing countries as a source of sustenance and livelihoods and as the storehouse of traditional health-care systems. The rights over and access to these resources and the income accrued from them also are crucial to the forest dwellers.

2. It is estimated that nearly 80 per cent of the people in developing countries rely on TMs.

---

[1] Visit: www.vindhyaherbals.com (Accessed in May 2015).
[2] Visit: www.oushadhi.org (Accessed in May 2015).

3. The commercialization of biological resources, especially of the MAPs, spans from village level to the international level as these products, apart from their household and health-care uses, find use in a diversity of industries, horticultural practices, crop protection and biotechnology. Many products enter the international trade, effecting enormous rise in the prices of the products as well as creating concerns for sustainability of biological resources.

4. The trade of MAP species within national boundaries (in India as well as in other tropical countries) is highly secretive and unorganized, while the international trade is controlled and regulated by various trade measures.

5. The tropical forests in the developing countries have traditionally been the major collection centres for biological resources, and in the Asian region, India, Indonesia, Malaysia, Thailand and China are major suppliers of these resources to the world market.

6. There is a clear indication that the general flow of trade in biological resources is directed mostly from the resource-rich and technology-poor 'South' to the biodiversity-poor and technology-rich 'North', and new market preferences in general for the natural products are much enhanced in the national and international trade.

7. Along with the issues related to the commercial use of these resources (including the access to the resources and the equitable benefit sharing), another important subject that draws much attention is the TK in possession of the indigenous and local communities.

8. The WTO and the enforcement of the Agreement on TRIPS, one of the most contested agreements under the umbrella of the WTO, have accentuated the debate over the commercial use of biological resources and the associated TK at global level.

Now let the functional tool be worked out that is useful for making maximum benefit for all stakeholders and putting together an equitable benefit sharing. Regarding an appropriate path of action that is advisable for this purpose, the best would be to follow what Charles R. McManis (2003) suggested: *Thinking globally, acting locally*. When this formula is put to use for drawing the pathway, aiming for an equitable sharing of the benefits accruing from the biological wealth of the 'South', it is recognized that one has to adopt the mechanism of benefit sharing (see Chapter 7) based on local customs, traditions and laws while keeping an eye on the international IPR regime.

If this mechanism—which basically is a commonality found in the case studies of partnership programmes—has to be adopted as the best practice, a need arises of having a well-equipped resource person who can act as a guide. The action of forging partnership contracts for access to the biological resources and equitable benefit sharing will need a professional. This hypothesis is described in the following section along with the desired qualifications this professional should have and the expectations from him or her.

While this process is adopted by the resource provider countries, step by step, it may be admitted that another action to be undertaken along is the continuation of the efforts at international level for seeking suitable changes in Article 27.3 of the Agreement on TRIPS with liberal interpretations of Articles 7 and 8 of the Agreement on TRIPS and with the support of provisions of the CBD. If put in the shape of recommendation, these three steps may be listed as follows:

Recommendation No. 1: Looking to the complexities involved in formulating the bioprospecting partnership agreements, a need arises to have professionals who have a sound knowledge about the ABS practices and regulations, and for this, it is recommended that the ABS should evolve into a new scientific-cum-legal discipline. The professionals well educated and trained in this discipline would prove to be of great help to all the stakeholders in the process.

Recommendation No. 2: There is no single formula like 'one size fits all', and treating every partnership programme as a different case, the mechanism as outlined in Chapter 7 has to be built up for each partnership programme. It is inevitable to go step by step to reach the final stage of agreements wherefrom the project would take off, and every step in this process is significant and needs to be treaded carefully.

Recommendation No. 3: During the recent past, significant developments have been taking place at international level under the umbrella of different UN organizations, and the resource-rich tropical countries have been dealing with the related issues in coherence. It may well be recommended that the tropical countries have to keep harping on for suitable changes in the

Agreement on TRIPS to make it compliant with the CBD, and they should continue their endeavour for rendering the international instruments legally binding, so that the anomalies in the sharing of benefits accruing from the commercial use of the components of biodiversity and the TK associated to these resources may be removed.

These recommendations are elaborated in the following sections in that order.

ABS: A New Scientific-cum-Legal Discipline (Recommendation No. 1)

It is well realized that the spade work to begin a partnership programme has to be on the lines narrated in Chapter 7 of this book. What the discussions, the successful partnership programmes and the efforts made at global level reveal is that an equitable sharing of the benefits that accrue from the commercial uses of biodiversity and the TK associated to the genetic resources cannot be denied to the indigenous and local communities. The question now is: Who will begin the spade work, how to start that work and what tools are needed for the purpose? It is also realized that there cannot be a rigid kind of framework, or a foolproof set of preferences, that would remove all the bottlenecks in the trade related to the commercial use of biodiversity. Therefore, instead of suggesting a formal model, going ahead with the concept of 'thinking globally, acting locally', it is suggested that the subject area of ABS should be evolved into an altogether new scientific-cum-legal discipline of study. The professionals who master in this discipline should delve into the eight specific issues described below, and they should be trained in a manner that they may be able to provide appropriate legal and professionally strong advice. In a nutshell, we may propound the ABS as a new professionalism, and this new scientific-cum-legal discipline should comprise chiefly the following major fields of knowledge:

1. **Laws related to biodiversity, patents and so on:** Sound knowledge about the biodiversity laws, the ABS regulations and patent laws and adequate exposure to the international legal

instruments related to biodiversity conservation and intergovernmental institutions. Adequate exposure to the national biodiversity action plans, the complexities in benefit sharing and a sound working knowledge about financial management. Basic understanding of laws, rules, company laws and by-laws. Laws governing the cooperative societies.

2. **Entrepreneurship and industrial production:** Scope of industrial development and future outlook, chiefly in the areas of pharmaceuticals and health care, agriculture, horticulture and rural livelihoods. Study of home-based as well as small- and medium-size enterprises, elements of bio-trade and pre-processing techniques, supply chain, value addition and other components of bioprospecting partnerships involving non-monetary steps in sharing of benefits.

3. **Nature sciences:** Basics of biodiversity (within species, between species and ecosystem biodiversity), biodiversity status and health and sustainable use of target species. Basics of ecology, forestry, taxonomy, ethnobotany and phytochemistry.

4. **Social sciences:** Study of socio-economic conditions and other local factors that in general have greatest influences including livelihood aspects of indigenous and local communities. Knowledge about local institutions, their roles and responsibilities, awareness about local interest groups and local NGOs, awareness about general ethics, social customs and traditions, local language and demography, prevalent cooperative and other socio-economic structures and framework. Rights of resource providers and users, PIC and evaluating bargaining powers of various groups.

5. **Stakeholder analysis:** Identification of on-site and off-site stakeholders, stakeholder analysis to evaluate their power and influence and working out the options in order to enforce future involvement of different interest groups. Competencies in eliciting input from experts and socially effective local institutions, and opening dialogues between the TK-holders and resource users.

6. **Art of negotiation and effectuating proper participation:** Capabilities in negotiating with commercial enterprises including transnational corporations, art of negotiation with rival community organizations and local institutions, sound knowledge

about fundraising techniques and evaluating options of loan management. Concept about who actually represent the local people, community participation, exposure to local issues, problems and tribulations. RRA and participatory approach aiming at an active participation of all, art of organizing seminars and meetings for involvement of stakeholders and inter-sectoral coordination between different government departments.

7. **Project formulation, approval, monitoring and evaluation:** Principles of project concept and planning, project identification and design, making strategic choices and planning exercise, developing work plan, monitoring and evaluating framework, data collection, analysis and reporting, preparing a timeline of development projects, performance indicators, good project planning and resource planning. Basics of actuarial science for improved financial decision-making.

8. **Conflict resolution:** Ability to foresee internal conflicts, dispute settlement and mainstreaming of local groups.

While the ABS professional should have long-term outlook and help in drawing future technological development framework, he should be kept out of the project once it takes off in order that every institution and organ of the partnership is able to work on its own. The concept of ABS as a new scientific-cum-legal discipline is based primarily on the notion that professional expertise may be able to manoeuvre different interest groups beginning from the planning stage and up to the project take-off stage, while the principal investigator of the project is somebody who is able to provide leadership to the whole gamut of the partnership programme. An ABS professional or group of ABS professionals may be assigned the task of preparing proposals for the project envisaged, facilitate the drafting of various multilateral agreements and by-laws, and extend all help in the processing of relevant proposals for approval as required under the ABS regulations or local biodiversity laws. To sum up, the ABS professional has to win the faith of all, including the on-site and off-site stakeholders, and in a way, he is an advocate for both the parties and he or she has to work on the principles of win-win.

The scheme of themes propounded under the scientific-cum-legal discipline of ABS is given in the ellipses serialized from 1 to 8 in Figure 9.1. To sum up, it may be stated that the ABS professional is

Figure 9.1:
ABS as a new scientific-cum-legal discipline

1. Laws related to biodiversity, patents, etc.
2. Entrepreneurship and industrial production
3. Nature sciences
4. Social sciences
9. Programme take-off and exit of the ABS professional

Access and Benefit Sharing:

A New Scientific-cum-Legal Discipline

5. Stakeholder analysis
6. Art of negotiation and effectuating proper participation
7. Project formulation, approval, monitoring and evaluation
8. Conflict resolution

The professional who is expert in ABS should be out of the scene, leaving behind a sound long-term outlook, once the programme takes off.

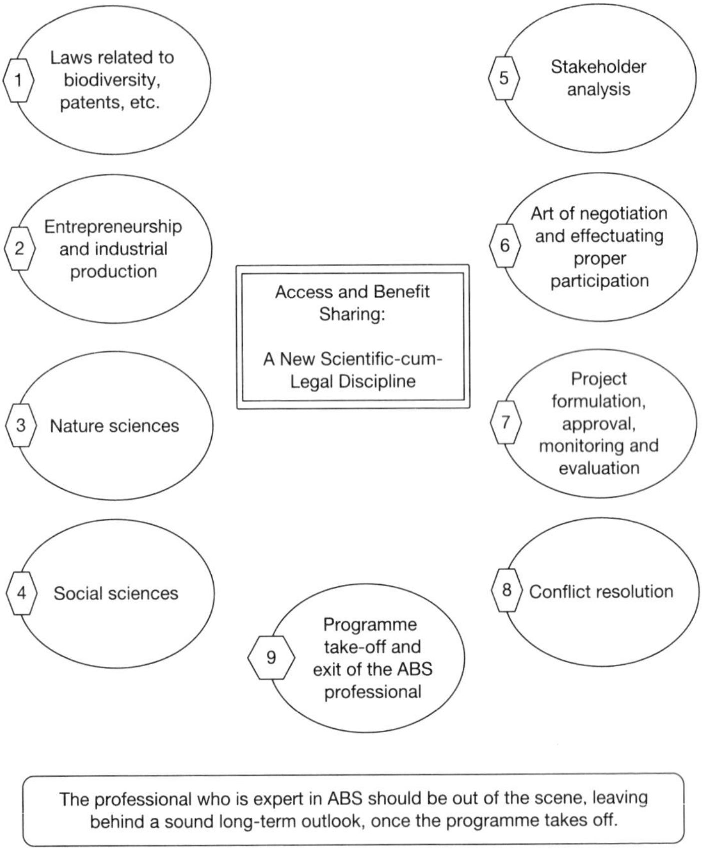

not expected to be a botanist or a taxonomist. It is not necessary for him to be a law graduate or a patent attorney. Of course, he or she should have sound primary knowledge in all these fields, and besides, he should have sound knowledge about what is a natural resource and how best it can be managed, and his field of knowledge should be across national boundaries—he is also expected to know the legal provisions in other parts of the world. His test of competency will be how adept he is in dealing with experts from quite different fields of knowledge. If we draw an analogy, an ABS professional is the one who is employed to prepare applications for various kinds of permission to make commercial use of genetic resources, just like a patent attorney who is engaged to facilitate the filing and grant of patents.

The last stage (ellipse no. 9) as shown in Figure 9.1 is the stage when the programme takes off, and that is the stage when the ABS professional has to leave the scene. The profile of the subject areas inscribed above in which the ABS professional has to be trained is only indicative, and depending on the institution or the level of degree or diploma (business management or like) to be awarded, suitable changes may be made in the profile of subject areas.

The Best Course of Action for ABS (Recommendation No. 2)

Subsequent to the recommendation that the ABS should be evolved as a new scientific-cum-legal discipline with professionals educated suitably to assist in forging appropriate partnership programmes, looking to the varied nature of these programmes, particularly with respect to the size and extent of the projects, it is appropriate to follow the mechanism as suggested in Chapter 7 for an equitable benefit sharing. The mechanism suggested is basically a denotation of the commonality found in the case studies reviewed in Chapter 6. The salient features of the mechanism of benefit sharing may be precised as follows:

1. It is essential to have the project size well defined and also the programme area well delineated, once the indigenous and local communities that the project would comprise are specified. The stakeholder groups involved have the choice of deciding the basic objectives to be achieved, and in continuation, the extent of the project area as well as the desired magnitude of finance has to be decided. The geographical region is delineated in a way that it covers the ecological region wherein the biological resources originate, and the exercise is done keeping in view the socio-economic conditions as well as cultural traditions of the local community.

2. The exercise of identifying and defining the various groups who matter most as the on-site stakeholders should be taken up with accuracy, and the major goals of the programme should be closely linked to these groups.

3. Deciding about the actual partners in various agreements or the identification of stakeholders involved is the next important task. It is essential to qualify the different categories of stakeholders, as different stakeholders have different goals often leading to conflict with one another. The process of identifying the potential partners from the list of off-site stakeholders is also crucial as even they have decisive roles to play. Simultaneously, stakeholder analysis is an important step, and what is needed is an approach to ensure stakeholder participation with opportunity to influence the decision-making at local level under given situations. Concurrently, involving women as a group among the on-site stakeholders is also pertinent because of their significant influences.

4. A broad framework of the entire programme including the goal setting and the time frame is required to channel the benefits accruing from the biological resources. The enormity of the programme and the complexity of various agreements that are of the essence, involving multiple partners who may have mega-influences, need tough deliberations. The process of enlisting all future activities takes time, sometimes several months, and finally the MOU, one of the main steps, is signed. The truth is that the process of drug discovery and research on endemic diseases may take several years, and accordingly, the ecological conditions have to be the focal point in this exercise.

5. An inventory of the biological resources available in the project area needs to be prepared with reliable information about the resources, including the distribution of different species over the landscape and the sources of origin of endemic species. Linking the resources with the local clan of the community with respect to the associated TK and safeguarding the representative samples and their traditional uses are relevant, so is the impact of commercial-scale harvesting of the chosen species. Similarly, there is the need for monitoring of the conservation status of species, for collection of the samples (specimen collection) and

also the sustainability issues that are of concern. Training the personnel deployed to prepare the inventory, appropriate mode of documentation and so on are also vital in the process.

6. The basic process of formulating the partnership agreements starts after an MOU has been arrived at. At this stage, the on-site stakeholders will need help in learning the art of negotiation including a lesson on how to use their bargaining power effectively under the given scenario, and the very process of formulating the eventual agreements between various partners of the project will begin. The agreements in most of the cases may be multilateral and the goals and objectives included in the MOU will have to be followed to prepare the first drafts. Subsequently, each partner may get scrutinized the draft agreements by suitable experts, and they may incorporate changes to eliminate probable flaws and inequalities. The process will also ensure that no genuine benefit claimer is excluded from consideration in the sharing of benefits accruing from the biological resources or the TK associated. The collaborators need to converse effectively to delve on all these issues. The different types of benefits and their method of equitable sharing (see Chapter 7) will also be a part of these agreements. A capacity-building part to coordinate the conservation and sustainable utilization of biodiversity should also be stipulated in the agreements.

7. As elaborated in Chapter 7, the mechanism of trust fund is most common and probably the best method for channelling the inflow of funds, an effective control on the expenditure part, particularly for the socio-economic developmental works for the indigenous and local communities, and also the outflow of the cash incentives and the benefits to various groups and individuals. The trust fund should be well planned and its constitution should be written and the governing bodies need to be formed keeping the local laws, customs and traditions under consideration.

In addition to the above seven points chosen as the best course of action for access to the biological resources of the 'South' and an equitable sharing of the benefits accruing from these resources and the TK associated to them, though the intricacies in this arrangement have already been explained in Chapter 7, the following few precautions are essential to be followed when we apply the formula of 'thinking globally, acting locally' in taking up any new biodiversity prospecting project:

1. *Every partnership programme should be treated as a different case.* Highly technical, complex and legal bioprospecting partnerships are being worked out now, engaging numerous institutions after tough negotiations often under mega-influences of global scale. Multilateral and long-term partnership programmes extending over a region, with adequate flexibility for writing different subprogrammes designed to match local variations and needs, and specific agreements for each subprogramme, cannot be replicable in toto. It is thus clear that every proposed partnership programme should be treated as a different case and each project should be studied in the light of its specific situations and should be treated afresh for writing an altogether new partnership programme, whether small or large. Also, the constitution of trust funds and other related legal formulations should be based on the local laws and regulations.

2. *Capacity building of indigenous and local communities should be an intrinsic part of all partnership projects.* An important component of the upfront payments that are built in any partnership programme is in the shape of capacity building of the indigenous and local communities. It includes specialized training in species identification and in the screening for pharmaceutical uses that may bring about skill improvement among local communities. These may include basic technological skills in extraction chemistry, chemical analysis and purification of samples. Other areas can be training in database management, equipment for sample preparation and storage and training in the technology of drug discovery, natural products and biochemistry research as parts of a larger programme. It sometimes includes development of source country infrastructure (comprising vehicles, laboratory improvement and community health care) and institutional partnerships for biodiversity management. It is recommended that capacity building of the

indigenous and local communities should be incorporated in all the partnership programmes in order to ensure future prospects of MAP species and to secure sustainable biodiversity management practices.

3. *Involvement of various sectors and the government should be ascertained.* As the financial arrangements in any programme bear more authenticity if the government is involved, as has been observed in many bioprospecting case studies (see Chapter 6), especially in the grant cases, it is imperative to involve the government as one of the active and constantly involved partners. The conclusions drawn in a range of the case studies point out that it should be ensured that no important sector (related government departments, public or corporate sector pharmaceutical enterprises, NGOs, international funding agencies, etc. besides the on-site stakeholders) is left out in large-size partnership programmes. It is suggested that various sectors that have significance should be involved in the bioprospecting programmes, and they should have proper communication, honesty and mutual trust. It is also proposed that in order to bear more authenticity, the involvement of the on-site stakeholders and the government should be ensured from the very beginning stage.

4. *Establishment of trust funds should be an essentiality.* Establishment of the trust fund is a key component of benefit sharing arrangements involving the commercial uses of genetic resources. Whatever may be the original source of the fund flow, and whatsoever may be the mode or channel of flow of benefits, the fund flow emanating from the royalty through the distribution mechanism and all other monetary benefits ought to be managed by an elected board. The system avoids the drawback that direct cash payments have, and this feature ostensibly makes the trust fund model highly acceptable. It can provide immediate benefits to on-site stakeholders as it can be initiated with the advance payments received as the compensation for TK held by the local communities. In a nutshell, the trust fund institutionalizes the benefit sharing mechanism and is recommended as the most important tool for equitable commercialization of genetic resource base.

Seeking Support of the International Instruments (Recommendation No. 3)

The analysis of different issues and the review of case studies in the context of the commercial use of biological diversity corroborate the dependence of the people of the countries of the biodiversity-rich and technology-poor 'South' on the components of biodiversity. The products of biodiversity in these countries are gathered and sold by the forest dwellers and rural people to the local traders in either raw form or dried and semi-processed form for monetary gain. Compared to the products gathered by the rural people for self-consumption or for making artefacts to earn their livelihood, the products gathered for monetary gain have greater commercial value and scope of bio-logical prospecting, comprising most of the MAPs used in manufac-turing industries. It has also been noticed that these manufacturing industries are situated in the biodiversity-poor and technology-rich countries of the 'North' and draw advantage from the TK associated to the biological resources, mostly without sharing of the benefits accruing from the commercial use of biological diversity.

It was also observed in the previous chapters how the indigenous and local communities are deprived of the larger share of the benefits that could accrue to them because of the limitations in the present IPR regime. The international legal instruments available today are not favourable to the protection of the TK, and instead, these instru-ments allow free access to the biological resources of the tropical countries. Simultaneously, it is highlighted that there are continual debates and deliberations going on at the international forums for incorporating suitable changes in the Agreement on TRIPS so that it complies with the provisions of the CBD, and there exist several trea-ties and institutions today that are supportive to the interests of the tropical countries.

What is required now is to aim for robust international legally binding instruments that would be helpful in the protection of TK and in drawing equitable benefits from the biological wealth of the tropical countries. To fulfil this aim, the biodiversity-rich tropical countries of the 'South' should seek support of the new developments happening at the international forum. Though the biodiversity laws and the ABS regulations of these countries may prove to be help-ful in ensuring an equitable sharing of benefits, it is still difficult to safeguard their interests in the present international IPR regime. For example, India's Biological Diversity Act of 2002 has adequate measures for regulating the ABS issues, yet the law is not applicable beyond the national boundaries and therefore the dependence remains on other alternatives, such as the TKDL if the misappropriation of the TK has to be prevented.

It would also be pertinent to evaluate the efforts made so far within the purview of a number of institutions and international treaties dealing with the prospects of commercial use of the components of biological diversity. In addition to the CBD and its supplementary agreements pointed out in the beginning of this chapter, and the several national laws and regulations related to the ABS, today we have the ITPGR for Food and Agriculture of the FAO, the International Union for the Protection of New Varieties of Plants (UPOV), the CITES and the Doha Declaration on the TRIPS Agreement and Public Health (WHO, 2001b). These instruments allude to the various aspects related to the biodiversity conservation. However, there is not so direct linkage between these instruments and the intricacies involved in ABS. Rather, there are many other parallel developments reshaping the issues of ABS internationally. These developments are critically examined in the following paragraphs.

Apart from the international framework mentioned above, the initiatives of the WIPO[3] also are of substantive value and significance. Being a global forum for IP services, policy, information and cooperation, the WIPO delivers global services for protecting IP and offers an open-for-the-world opportunity to learn all about IP and its protection, very much including future scope even for the protection of TK. In fact, negotiations are currently underway in the WIPO Intergovernmental Committee on Intellectual Property and Genetic Resources, Traditional Knowledge and Folklore (IGC) towards the development of an international legal instrument or instruments for the effective protection of traditional cultural expressions and TK, and to address the IP aspects of access to and benefit sharing in genetic resources.

Surely, with more awareness among the like-minded developing countries and with the involvement of institutions such as WIPO

---

[3] A self-funding agency of the UN with 187 member states and headquarters at Geneva, the WIPO has its mandate, governing bodies and procedures set out in the WIPO Convention. Website: www.wipo.int (Accessed in May 2015).

and the SCBD, the issues related to ABS have to be given a concrete shape before long. To put it more plainly, the Agreement on TRIPS needs to be 'corrected' suitably and accredited also by the developed countries. The attempts made so far to pursue the review of the Agreement on TRIPS for incorporating appropriate amendments with the support of the CBD are getting strengthened with the new developments taking shape at international forums. If we review the beginning of this phenomenon with the Doha Declaration on the TRIPS Agreement and Public Health (WHO, 2001b), which responds to the concerns of developing countries about the obstacles they faced when seeking to implement the measures to promote access to affordable medicines in the interest of public health in general,[4] and we look at the Aichi Biodiversity Targets of 2010 (the subject discussed in the following paragraphs), we find that the day may not be too far when, conforming to the objectives of the CBD, suitable amendments in the Agreement on TRIPS would be accepted as a final point. Paragraph 19 of the Doha Ministerial Declaration of 2001 (WTO, 2001) has broadened this discussion, saying that the Council for TRIPS should also look at the relationship between the Agreement on TRIPS and the CBD, the protection of TK and folklore (see Box 9.3).

At this juncture, let us examine how helpful are the endorsements of the CBD to the tropical countries. The concern and the interests of the tropical countries were endorsed greatly by the CBD, as the three major goals of the CBD (conservation of biological diversity, sustainable use of its components and fair and equitable sharing of the benefits accruing from the use of genetic resources) closely relate to the biological wealth of the tropical countries. Though the CBD by itself does not provide legal protection to the TK-holders over their knowledge or genetic resources and it leaves benefit sharing policy to be defined in national laws, it acknowledges the vital role of TK associated with the biological resources and endorses the access to these resources on MAT. One very significant development within the scope of CBD was the adoption of Strategic Plan for Biodiversity 2011–2020 in the 10th meeting of its COP held in Nagoya in October 2010. The mission of this plan, going beyond the three major goals

---

[4] *Source*: http://www.who.int/medicines/areas/policy/doha_declaration/en (Accessed in September 2014).

Box 9.3:
Doha Ministerial Declaration 2001 (Paragraph 19)

> We instruct the Council for TRIPS, in pursuing its work programme including under the review of Article 27.3(b), the review of the implementation of the TRIPS Agreement under Article 71.1 and the work foreseen pursuant to paragraph 12 of this Declaration, to examine, inter alia, the relationship between the TRIPS Agreement and the CBD, the protection of TK and folklore, and other relevant new developments raised by Members pursuant to Article 71.1. In undertaking this work, the TRIPS Council shall be guided by the objectives and principles set out in Articles 7 and 8 of the TRIPS Agreement and shall take fully into account the development dimension.
>
> **Source:** www.wto.org/english/thewto_e/minist_e/min01_e/mindecl_e.pdf (Accessed in September 2014).

of the CBD, aims at reduced pressure on biodiversity, restoration of ecosystem services, providing adequate financial resources, enhancing capacities and mainstreaming biodiversity issues and values.

The Strategic Plan is enriched not only with policy guidelines but also with clear-cut action points. It has 20 specific targets catalogued as the Aichi Biodiversity Targets under five strategic goals.[5] In the context of ABS and related issues, the following four specific targets under the strategic goals of enhancing 'the benefits to all from biodiversity (Goal D)' and 'implementation through participatory planning, knowledge management and capacity building (Goal E)' are an excellent illustration of the journey of CBD beyond 2010:

- By 2015, the Nagoya Protocol on Access to Genetic Resources and the Fair and Equitable Sharing of Benefits Arising from Their Utilization (see Box 9.4) is in force and operational, consistent with national legislation (Target 16).

---

[5] Details available on the CBD website: www.cbd.int (Accessed in May 2015).

Box 9.4:
Nagoya Protocol

The **Nagoya Protocol on Access to Genetic Resources and the Fair and Equitable Sharing of Benefits Arising from Their Utilization to the CBD** was adopted in 2010 during the 10th meeting of the COP to the CBD in Nagoya, Japan, in order to effectively implement Articles 15 (*Access to Genetic Resources*) and 8(j) (*Equitable sharing of the benefits of traditional knowledge*) of the CBD. The Nagoya Protocol was the result of six years of negotiations of the Ad Hoc Open-ended Working Group on ABS in response to the WSSD held at Johannesburg in 2002. The Protocol paves the way for commercial practices of the CBD's objective of fair and equitable sharing of the benefits of utilization of genetic resources by providing a strong basis for internationally accepted legal certainty and transparency for both—the resource providers and the resource users. The Protocol, which brought in the concept of benefit sharing under international law for the first time by providing legally binding rules on the compliance of ABS, has received initial ratifications almost entirely from the developing countries. It entered into force on 12 October 2014.

The Nagoya Protocol takes the objectives of the CBD further ahead of the Bonn Guidelines, with emphasis on:

- Legal certainty and clarity of ABS legislation;
- Regulatory requirements for accessing genetic resources and appropriate procedures for PIC involving the indigenous and local communities;
- Clear rules and procedures for MAT, including suitable dispute settlement clause and terms on benefit sharing, including on the IPR and on subsequent third-party use;
- Designating a national focal point and one or more competent national authorities on ABS;
- Addressing situations of non-compliance with appropriate legislative, administrative or policy measures; and
- Promoting access to technology through transfer of technology to developing countries, least developed countries and small island-developing States in order to enable the

development and strengthening of a viable technological base for the attainment of the objectives of the CBD.

The Protocol also stresses on the development of model contractual clauses for MAT, and it is supplemented with a diverse range of the monetary and non-monetary benefits that would accrue to the resource providers. The annexure of these benefits is quite similar to the annexure appended to the Bonn Guidelines.

Source: www.cbd.int/abs/doc/protocol/nagoya-protocol-en.pdf (Accessed in September 2014).

- By 2015, each Party has developed, adopted as a policy instrument, and has commenced implementing an effective, participatory and updated national biodiversity strategy and action plan (Target 17).
- By 2020, the TK, innovations and practices of indigenous and local communities relevant for the conservation and sustainable use of biodiversity, and their customary use of biological resources, are respected, subject to national legislation and relevant international obligations, and fully integrated and reflected in the implementation of the Convention with the full and effective participation of indigenous and local communities, at all relevant levels (Target 18).
- By 2020, at the latest, the mobilization of financial resources for effectively implementing the Strategic Plan for Biodiversity 2011–2020 from all sources, and in accordance with the consolidated and agreed process in the Strategy for Resource Mobilization should increase substantially from the current levels (Target 20).

After having discussed the state of affairs dominated by the constituents and the components of the CBD, it would be pertinent to have one final look at the most important happening, the advent of the WTO and especially the Agreement on TRIPS, before looking for the alternatives that the biodiversity-rich developing countries apparently vouch for the equitable sharing of benefits of the biological diversity of the world.

The arrival of the WTO in 1995 has been the most significant happening in the area of the present-day international trade regime. Its predecessor, the GATT, attracted little attention as regards the conservation and commercial use of the biological resources of the 'South'. Widening the scope, duration and strength of patent protection, the Agreement on TRIPS is now the main instrument in the global patent system, viewed as having implications over the commercial uses of the genetic resources and associated TK of the tropical countries. The progress of biotechnology involving natural products is an issue contemplated to be more imposing in this context, and more recently, vast experiences have been gained also in the benefit sharing arrangements related to bioprospecting. Still a majority of the WTO membership conceives the Agreement on TRIPS to be predominantly for the benefit of industrialized countries.

The Agreement on TRIPS provides that patents shall be enjoyable without any distinction whether products are imported or locally produced and extends patent protection to microorganisms, non-biological and microbiological processes and even the plant varieties. Regarding the infringement of IPRs, it states that the burden of proof shall be on the defendant, that member countries shall enforce effective action against any act of infringement of IPRs, and also that measures should be taken to avert the creation of barriers to legitimate trade. Another peculiarity is that even the concept of trade secrets (though not defined in the Agreement) is protected under the general term 'undisclosed information' (Article 39) and it is left to the individual members as to how to provide such protection. The term 'the submission of... other data the origination of which involves a considerable effort shall protect such data against unfair commercial use' in Article 39.3 of the Agreement, according to Charles R. McManis (2004), appears broad enough to cover any TK that the WTO members might require to be disclosed to a government agency as a condition of approving the marketing of pharmaceutical or agricultural products that utilize new chemical entities, regardless of the reason for the required disclosure of such information.

These consequences make it obligatory to devise a mechanism to protect the body of TK and to ensure equitable sharing of benefits that accrue from the realization of its commercial value. The tropical countries may also draw strength from Article 8 (*Principles*) of the Agreement on TRIPS, which states that (a) the members may adopt measures necessary to protect public health and nutrition, and to promote the public interest in sectors of vital importance to their socio-economic and technological development, and that (b) appropriate measures may be needed to prevent the abuse of IPRs by right holders or the resort to practices that unreasonably restrain trade or adversely affect the international transfer of technology.

On the part of tropical countries, in the perspective of seeking support of the international instruments, it is stressed that there is a need for a symbiosis between the Agreement on TRIPS and the CBD, even if their objectives point to different overall goals, in order to meet the objective of fair and equitable sharing of the benefits. The CBD makes a striking shift from unfair exploitation to a more balanced specification that requires a prudent system of exchange of biological resources in lieu of the transfer of manufacturing technologies. Drawing support from the CBD, it has been suggested that the Agreement on TRIPS should be amended to secure that whenever a patent is granted to a product or process based on biologically derived inventions that incorporate TK, the applicant of such a patent must provide the following:

- The disclosure of the source and country of origin of the biological resource and the TK associated with it that was incorporated in the invention;
- The confirmation of PIC to prove that the materials were accessed legally and the TK was used with the consent of the respective holders of the TK; and
- There exist MAT for fair and equitable benefit sharing.

In order to meet the above requirements, it is recommended that Article 29 (*Conditions on Patent Applicants*) of the Agreement on TRIPS should be suitably amended and the rights of farmers and local communities must be reflected. In this respect, with support from like-minded countries, India has been advocating in its submissions to the review meetings of the Council for TRIPS for a proper linkage between the CBD and the WTO. It would be noteworthy in this context that the three points of concern mentioned above (the disclosure clause, the provision for PIC and MAT for benefit sharing) have become part of one or the other biodiversity laws of many tropical countries. As the Articles 7 and 8 of the Agreement on TRIPS mention that IPRs should contribute to social and economic welfare and stipulate prevention of abuse of IPR that will restrain trade or adversely affect international transfer of technology, what is required now is a liberal interpretation of these Articles.

In this context, the good thing that has happened is that many biodiversity-rich tropical countries have formulated a range of legislations for ABS and also for the conservation of biodiversity and the protection of TK associated with their genetic resources (see Chapter 5). India ranks among the foremost countries that have evolved legislation for ABS arrangements, for the conservation of biodiversity, and also for the protection of TK held by the indigenous and local communities. India's Biological Diversity Act of 2002 is the instrument that serves this purpose with its aims and objectives agreeing wholly with those of the CBD. The provisions of this law were compared with the provisions made under the Agreement on TRIPS and the CBD in Chapter 5, and how this legislation does help in actualization of the benefits accrued from the utilization of genetic resources and the associated TK may be summed up in short in the following list:

1. **Enforcement of IPRs:** India's Biological Diversity Act of 2002 stipulates for permission, in India as well as outside, for IPRs related to the biological resource-based inventions. An applicant applying for patents, or for obtaining India's bioresource or associated knowledge, shall have to obtain permission from the NBA for the purposes of research, commercial utilization or for the transfer of research results. Such permission shall be on specified terms and conditions including royalty imposition. The NBA may also adopt measures for challenging the grant of IPRs abroad on the biological resource obtained from India or associated TK.

2. **Protection of TK:** On the recommendation of the NBA, the GOI shall undertake measures for the protection of indigenous knowledge relating to biodiversity, including measures for registration of such knowledge through steps including sui generis system. At local level also, the local bodies shall have to constitute BMC for promoting conservation, sustainable use and documentation of biodiversity including chronicling of associated TK.

3. **Access to genetic resources:** For access to the biological resources of the country and associated TK to any non-Indian

individual or corporate body having non-Indian participation, the Act provides for prior approval from the NBA for the purposes of survey, utilization, research or for any commercial use. The results of such research works also cannot be transferred to non-Indian individuals or corporate bodies having non-Indian participation without prior approval.

4. **Mechanism for benefit sharing:** The Act has some prerequisites with respect to the benefit sharing arrangements accruing from the use of biological resources, which are summarized as follows:

- NBA while granting approval may impose fee or royalty from commercial use of IPR (Section 6.2).

- No person (excluding local people and communities, growers and cultivators of biodiversity and vaids and hakims) to obtain biological resource for commercial use or biosurvey except after giving prior intimation to the SBB concerned (Section 7).

- Benefit sharing through joint ownership of IPRs, transfer of technology, location of production or R&D units and setting up of 'venture capital fund' (Section 21).

- Benefit sharing money to be kept in the NBF (constituted under Section 27 of the Act), to be used for channelling benefits to the benefit claimers, conservation of biological resources and socio-economic development (Section 27.2).

- BMCs may levy charges by way of collecting fees from any person for accessing or collecting any bioresource for commercial use (Section 41.3).

Though these provisions cannot be categorized as the best solution, these are fairly inventive as regards the equitable sharing of benefits accruing from the biological resources and the associated TK, and the provisions may offer great strength to the indigenous and local communities as regards the ABS. The stipulations that form the genesis of most of the biodiversity laws or ABS regulations of biodiversity-rich tropical countries in consonance with the provisions of the CBD may not be adequate in removing all the bottlenecks in the trade and commercial uses of the genetic resources of the respective nations. Yet it may be concluded that they should prove of great help in chalking out more viable arrangements for benefit sharing through contractual mechanism to be adopted in carefully negotiated bioprospecting partnership programmes.

One drawback that is apprehended in many legislations in these countries is that the central character or the nodal role to be played by the national-level institutions is likely to complicate the process of granting approval and permission, and this situation may invite flak from the international trade regime as the Agreement on TRIPS (Article 41.1) states that the enforcement framework of the IPRs should not by itself become barrier to legitimate trade. In other words, while the genetic resources are mostly confined to smaller physiographical regions and the laws are by and large having a general and extensive character spanning over larger regions across cultural boundaries and varying customary laws, it is surely advisable to have the contractual agreements based on local traditional and cultural practices and local customary laws.

It would be far better that the central character of the national-level institutions is not authoritative and the local institutions are appropriately empowered to take their own decisions. In certain respects, India's Biological Diversity Act of 2002 fulfils this criteria quite effectively.[6] Another noteworthy point is that already some countries such as India, Costa Rica, South Africa and the Philippines have passed enabling legislation for regulating the access to genetic resources. In the absence of a transparent system for obtaining PIC, frequently through certain process of permission, the system in effect for sharing benefits can be complex and difficult.

Realistically, the developing countries will need to align with the WTO-mandated IPR regime as spelled out by the Agreement on TRIPS, and, aiming to safeguard their interests, they would have to continue trying for a symbiosis between the Agreement on TRIPS and the CBD. These countries have been participating in the international-level debates in coherence, sitting together with like-minded countries, and fortunately, significant developments have been taking place within the ambit of UN organizations. If the recent initiatives undertaken at the various sessions of the IGC under the WIPO

---

[6] As per the NBA guidelines, on the recommendation of the SBB, the state government shall designate Nodal Officers for each of the districts to oversee matters at that level, and the BMC formations can be mediated through institutions or civil society organizations or Technical Support Groups (NBA, 2013a).

(expressed in more detail in the following section) are any indication, the tropical countries should be hopeful of winning necessary support from the international organizations and may have useful legally binding international instruments for the protection of their biological wealth.

There appear to be some concrete results in this direction, and after having discussed the options available to the tropical countries in the previously discussed three broad recommendations, we may go over some recent developments taking place at the UN forums, as a more encouraging scenario may emerge out of this exercise.

Some Recent Developments

All through this process, the Agreement on TRIPS is the focal point with respect to the biodiversity-rich tropical countries, and the recent years have witnessed, especially within the field of IGC, a combined genuine interest of the WIPO member states to engage in various sessions to find a fair solution to these concerns. Though the Agreement on TRIPS has no specific provisions on the issue of TK, the periodic sessions of the IGC have been having long and elaborate debates over these concerns. For example, the 26th session of the IGC held in Geneva in February 2014, participated by the member states including Brazil, India, China, South Africa, Bhutan, Bangladesh, Indonesia, Malaysia, Costa Rica, Namibia, EU, Russian Federation, UK, USA, Japan, Canada and Australia and many NGOs as well as intergovernmental organizations (FAO, UPOV, UNCTAD, UNESCO, UNEP, WHO, etc.) as observers, has the following salient points in its summation (WIPO, 2014b):

1. There was a general view among all delegations that the misappropriation of genetic resources and TK that can be deemed to be legitimately held is not acceptable. The matter of defining what is 'legitimately held' and what would constitute 'misappropriation' thereof needs further substantive discussion. No delegation expressed support for the view that misappropriation should be condoned or should be deemed unworthy of attention in this process.

2. A significant goal of the IGC process should be to find ways of assisting in preventing the erroneous granting of patents, and the instruments being pursued should extend to different forms of IPR.

3. There was no significant opposition to the possible benefit of using databases to help in checking whether a patent application should be denied on the basis that it constitutes *prior art*. (However, it was also stressed that any database mechanism would need careful consideration and must have adequate safeguards, or else the establishment of databases could lead to the misuse of information.)

4. The engagement of industry with member states and the rest of the stakeholders was strongly encouraged within the IGC process.

Also stressing that the WIPO was dynamic and continuously delivering compared to other international organizations, there was a general view that the elements such as ABS requirements should be tackled by domestic laws. Some of the other noteworthy reiterations made by the resource-rich countries and NGOs in the 26th session of the IGC were as follows:

- A provision for the establishment of a due diligence system to ascertain that protected genetic resources were accessed in accordance with applicable ABS requirements.

- It was necessary for the IGC to explore the possibility of establishing an effective mandatory disclosure requirement, which would protect genetic resources, their derivatives and associated TK against misappropriation and would prevent the granting of erroneous patents. (*However, many countries already have formulated ABS regulations that make the disclosure requirement mandatory as has been made in India's Biological Diversity Act 2002.*)

- In addition to the core issue that needs to be addressed (the issue of disclosure), the importance of elaborating a rights-based international instrument for the protection of genetic resources and associated TK and acknowledging the urgency of addressing misappropriation with a clear focus on IP was highlighted.

These developments are a follow-up of the decision of the WIPO General Assembly held in 2013 to agree to renew the mandate of the

IGC, incorporating that the IGC would continue to expedite its work with open and full engagement, on text-based negotiations with the objective of reaching an agreement on an international legal instrument that will ensure the effective protection of genetic resources and TK (WIPO, 2013). Historically, it was the Doha Ministerial Declaration of 2001 (WTO, 2001) that commanded and authorized the Council for TRIPS to examine the issue of protection of TK associated with the genetic resources, and the concurrent discussions on the relationship between the CBD and the Agreement on TRIPS began much earlier under the built-in review provision of the latter's Article 27.3(b). Though the TK-holders of the biodiversity-rich 'South' have not seen even a single change so far in the Agreement on TRIPS, it is noticed that the international debate is showing a sign of convergence now.

An example of this convergence is evident in the context of the digital database of the TK system. During the 26th session of the IGC held in Geneva (3–7 February 2014), the document submitted by the delegations of Canada, Japan, the Republic of Korea and the USA consisted of a joint recommendation on the development and widespread use of elaborated database of the genetic resources in each country for the defensive protection of genetic resources and TK associated with genetic resources (WIPO, 2014a). Referring to it as a structure of the 'one-click database search system', the joint recommendation mentions that:

- Each participating WIPO member state will collect information on genetic resources and non-secret TK associated with genetic resources within its territory and store this information in its database.
- The 'one-click database search system' should be an operational and searchable database under the proposed system, to be in the possession of, and maintained by, each participating WIPO member state.
- The database will be composed of a WIPO portal site as well as databases of WIPO member states, which are linked to this portal site. The system would help examiners conduct searches more efficiently for *prior art* or reference material concerned with genetic resources and non-secret TK associated with genetic resources, while preventing inappropriate access to its contents by third parties.

The basic purpose of having a functional database search system is to address the issue of TK, and it was agreed in the 2013 WIPO General Assembly for the IGC to expedite its work with open and full engagement. It was also decided that the focus in the 2014–15 biennium would build on the existing work and use of all WIPO working documents, and that the WIPO General Assembly should specifically address the issue of TK in the 27th session of the IGC (WIPO, 2014c). The negotiations are ongoing under the umbrella of the WIPO presently towards the development of international legal instruments for the effective protection of TK and to address the IP aspects of ABS.

However, it is evident from the review of recent documents of the WIPO related to the main institutions dealing with legalities of TK that adoption of an international legal instrument that is acceptable to all is still too far. In the 27th session of IGC held in 2014, the Chinese delegation remarked (WIPO, 2014d: paragraph 64) that the basic articles still reflected divergent views despite the efforts made in the previous IGC sessions, and it was deplorable that IGC was still far from coming up with one or several international instruments. The Japanese delegation also observed (WIPO, 2014d: paragraph 65) that despite a long history of discussion, the IGC had not been able to find common ground on the fundamental issues yet, namely on policy objectives, guiding principles, subject matter of protection and beneficiaries. In the same session of IGC, the representative of an international NGO stated (WIPO, 2014d: paragraph 44) that 'even if traditional knowledge and traditional cultural expressions were erroneously placed in the public domain, indigenous peoples should still be the rights-holders and have the right to compensation'. It was also added that effective participation of indigenous peoples in the IGC process was indispensable for devising the international legal framework regarding the rights of these peoples.

Surely the way the delegates in the 27th session of IGC alluded to the term 'indigenous people' vis-à-vis the term used earlier (indigenous and local communities) leads to an emphasis on the rights of the TK-holders. This was more apparent from the observation made by Ms Jennifer Corpuz of the Philippines-based Tebtebba Foundation[7] (WIPO, 2014d: paragraph 160) that the CBD was now undergoing discussions to change the terminology from 'indigenous and local

---

[7] Indigenous Peoples' International Centre for Policy Research and Education.

communities' to 'indigenous peoples and local communities'. It only brings home the legitimacy of a more equitable sharing of the benefits accruing from the commercial uses of biodiversity vis-à-vis the TK-holders.

Though the clarity on this issue is more prominent when we pay particular attention to the observations of different delegations from the tropical world, the opinion of the delegation of the USA complicates it appallingly. The US delegation pointed out (WIPO, 2014d: paragraph 62) that TK and traditional cultural expressions were distinctly different subject matters, that TK was a specific type of knowledge, whereas the traditional cultural expressions were specific types of creative expressions, and in the context of an international legal instrument, it would be difficult to agree upon a precise definition of TK and traditional cultural expressions. Whatever may be the scenario today, in the present context, the US delegation at least showed a pointer that the beneficiaries of protection should be indigenous and local communities, those who generated, used, held and maintained TK and traditional cultural expressions, and rewarding the originators of the TK and traditional cultural expressions would help to incentivize the generation of new knowledge and the creation of new creative expressions. This is for sure a significant development, and these framework developments have also pointed out that the research works based on TK may benefit from patent protection, and it might be practicable that illegitimate patent rights are not granted over TK subject matter that is not a true invention.

The IGC was assigned with the task of submitting the texts for an international legal instrument that would include a consolidated document on IP and genetic resources and a set of Articles for the protection of TK as well as traditional cultural expressions. The final decisions would be expected to be reached at in the WIPO General Assembly, though it is a difficult task if we look at the ambiguities of the draft articles. Yet the policy objectives included in the draft are bound to provide eventually for effective protection to the genetic resources and TK. The General Assembly may finally consider the need for additional meetings and decide on convening a diplomatic conference to reach the finality. The policy objectives make it clear (WIPO, 2014e) that the international legal instrument should aim to:

- Provide indigenous and local communities with the appropriate measures to prevent the misappropriation of their TK, control ways in which their TK is used beyond the traditional and customary context, promote an equitable sharing of benefits arising from their use with PIC or approval and prevent the grant of erroneous patents over TK and associated genetic resources.

- Facilitate the development of national TK databases for the defensive protection of TK, and for transparency, certainty, conservation purposes and trans-boundary cooperation.
- Provide that the IP offices shall ensure that such information is maintained in confidence, except where the information is cited as *prior art* during the examination of a patent application.
- Ensure that appropriate enforcement procedures are made available by member states under their laws against the wilful infringement of the protection provided to TK or misuse of TK sufficient to constitute a deterrent to further infringements.
- Provide that the IP applications that concern any process or product that relates to or uses TK should include information on the country from which the applicant collected the genetic resource or received the TK associated, and also state whether PIC or approval and permission to access and use has been obtained.
- Exempt from the purview of this protection of TK the use for non-commercial purposes such as teaching, learning, preservation, research and presentation in archives, museums and libraries, its use in cultural institutions recognized under national laws and creation of an original work of authorship inspired by TK.

The broad policy also envisages that in instances where the same TK is found within the territory of more than one member state, or is shared by one or more indigenous and local communities in several member states, those member states should cooperate with the involvement of indigenous and local communities concerned with a view to implement the objectives of the international legal instruments. Incidentally, this broad policy is also part of the Nagoya Protocol (Article 11: *Trans-boundary Cooperation*) (see Box 9.5), and applies equally to the genetic resources and the TK associated with genetic resources when having trans-boundary locations.

Box 9.5:
Nagoya Protocol on Access and Benefit Sharing: Article 11 (Trans-boundary Cooperation)

1. In instances where the same genetic resources are found in situ within the territory of more than one Party, those Parties shall endeavour to cooperate, as appropriate, with the involvement of indigenous and local communities concerned, where applicable, with a view to implementing this Protocol.
2. Where the same TK associated with genetic resources is shared by one or more indigenous and local communities in several Parties, those Parties shall endeavour to cooperate, as appropriate, with the involvement of the indigenous and local communities concerned, with a view to implementing the objective of this Protocol.

Source: www.cbd.int/abs (Accessed in September 2014).

Evidently, the deliberations in various sessions of the IGC have been momentous, and all stakeholders have very high aspirations from the purpose the institution of WIPO has shown and the utility it has proved so far. It is also very heartening for the biodiversity-rich tropical countries that an annual series of training courses on IP is being jointly convened since as early as 2009 by the WIPO and the WTO for government officials from developing countries, least developed countries and countries with economies in transition, aiming to strengthen the capacity of participating countries to monitor and participate in international developments, and to make informed assessments of IP policy issues. The main objective of the series is to update participating government officials on the activities and instruments of the WIPO and the WTO and to provide a forum for an exchange of information and ideas between them and the two secretariats on these matters. The course addresses the topics such as an overview of international policy and law in IP; the negotiating background of the Agreement on TRIPS and an overview on subsequent developments; IP and economic development; the Agreement on TRIPS and public health; innovation and IP; IP and genetic resources, TK and folklore; the relationship between the Agreement on TRIPS and the CBD; and enforcement of IPR.

In conclusion, the main finding we come across is that the components of biodiversity are crucial to the forest dwellers and members of the indigenous and local communities in tropical countries as nearly 80 per cent of the population in these countries depend on biological resources for their primary needs of health care, nutrition and income generation. The commercial use of biodiversity of these countries is closely linked to the economic interests and well-being of the people inhabiting these countries, and therefore, it is recognized that the manufacturing industries that draw the benefits from access to these resources and the use of the TK associated must share the benefits drawn with these people. Certainly an equitable sharing of benefits has to include the members of indigenous and local communities, even though the present international IPR regime appears to be unsupportive in providing IP protection to the TK. Eventually the biodiversity-rich tropical countries have to resort to bioprospecting partnership programmes in order to avoid the anomalies in the benefit sharing process. A mechanism has been propounded in Chapter 7 as the best course of action for ABS that is basically a commonality found in the case studies of partnership programmes from different parts of the world. It is also proposed that for the sake of formulating partnership programmes, the ABS should be evolved into a new scientific-cum-legal discipline, and a team of ABS professionals might prove of great help in this regard. The developments of the last 10–15 years at international forums, particularly under the aegis of the WTO and the WIPO and with huge support from the CBD, transpire immense hope for a bright future of the commercial uses of biodiversity going along with an equitable benefit sharing.

References

AICEP (1994). Summary Report of the All India Coordinated Ethnobotany Project of Ministry of Environment & Forests, Government of India. *As cited by* Shankar, D. and B. Majumdar (1997).

Alexiades, Miguel N. and Sarah A. Laird (2002). Laying the foundation: Equitable biodiversity research relationships (Chapter 1). In: Laird, Sarah A. (ed.), *Biodiversity and Traditional Knowledge: Equitable Partnership in Practice*. Earthscan Publications, London, Sterling VA (USA), pp. 3–15.

Anuradha, R.V., B. Taneja and A. Kothari (2001). Experiences with Biodiversity Policy-Making and Community Registers in India: Participation in Access and Benefit Sharing Policy Case Study No. 3. Swiderska, K. (ed.), IIED, London, pp. 58. *Source*: www.cbd.int/financial/bensharing/india-abs.pdf (Accessed in May 2015).

Aponte, José C., Abraham J. Vaisberg, Rosario Rojas, Michel Sauvain, Walter H. Lewis, Gerardo Lamas, César Sarasara, Robert H. Gilman and Gerald B. Hammond (2009). A multipronged approach to the study of Peruvian ethnomedicinal plants: A legacy of the ICBG-Peru Project. *Journal of Natural Products*, 72(3): 524–526. *Source*: www.ncbi.nlm.nih.gov/pubmed (*abstract only*) and http://wenku.baidu.com/view (full text; both sites last accessed in May 2014).

Arvigo, R. and M. Balick (1998). *Rainforest Remedies: 100 Healing Herbs of Belize* (2nd Revised and Expanded Edition). Lotus Press, Twin Lakes, Wisconsin, USA.

Aylward, B. (1995). The role of plant screening and plant supply in biodiversity conservation, drug development and health care. In: Swanson, T.M. (ed.), *Intellectual Property Rights and Biodiversity Conservation: An Interdisciplinary Analysis of the Values of Medicinal Plants*. Cambridge University Press, Cambridge, USA, pp. 93–126.

Balakrishna, P. (1998). Convention on Biological Diversity, Intellectual Property Rights and Voluntary Codes of Conduct: Facilitating Access and Benefit Sharing. In: South and Southeast Asia Regional Workshop on Access to Genetic Resources and Traditional Knowledge, Chennai, 22–25 February. IUCN Regional Biodiversity Programme, Asia. *Source*: www.bklabs.htm (Accessed in September 2004).

Balick, M.J. (2003). Traditional knowledge: Lessons from the past, lessons for the future. Draft paper for discussion. *Source*: http://ls.wustl.edu/centeris/Confpapers/ (Accessed in January 2005).

———— (2006). Ethnomedicine: Ancient Wisdom and Modern Science. Interview by Bonnie Horrigan. *Explore* 2(3): 239–248. *Source*: www.nybg.org/science/scientist_ profile.php/Interview_with_Michael_Balick.pdf (Accessed in November 2014).

Balick, M.J. (2007). Traditional knowledge: Lessons from the past, lessons for the future (Chapter 19). In: McManis, Charles R. (ed.), *Biodiversity and the Law: Intellectual Property, Biotechnology and Traditional Knowledge*. Earthscan, London, pp. 280–296. *Source*: www.planta.cn/forum/files_planta/biodiversity_and_the_law_107.pdf (Accessed in November 2014).

Barrett, Scott (1994). The biodiversity supergame. *Environmental and Resource Economics* 4: 111–122. Centre for European Economic Research, Mannheim, Germany. *Source*: www.link.springer.com/journal/10640/4/1/page/1 (Accessed in May 2015).

Bhojvaid, P.P. (2003). Medicinal plants based forest management: Problems and prospects. *Indian Forester* (Indian Medicinal and Aromatic Plants Special – I) 129(1): 25–36.

BIO (2013). Proposal for Reform of Brazil's Bioprospecting and Genetic Resources Regulations. Biotechnology Industry Organisation. November 18. *Source*: www.bio.org/.../BIO_Brazil_Bioprospecting_&_Genetic_Resources_FINAL.pdf (Accessed in June 2014).

Boutelle, J. (2004). Understanding Organizational Stakeholders for Design Success. *Source*: www.boxesandarrows.com/archives/ (Accessed in October 2012).

Brown, K. (1995). Medicinal plants, indigenous medicine and conservation of biodiversity in Ghana. In: Swanson, T.M. (ed.), *Intellectual Property Rights and Biodiversity Conservation: An Interdisciplinary Analysis of the Values of Medicinal Plants*. Cambridge University Press, Cambridge, USA, pp. 201–231.

Bryson, J.M. (2004). What to do when stakeholders matter: Stakeholder identification and analysis techniques. *Public Management Review* 6(1): 21–53. *Source*: www.hhh.umn.edu/people/jmbryson/pdf/stakeholder_identification_analysis_techniques.pdf (Accessed in May 2015).

Campbell, L.M. (2002). Conservation narratives in Costa Rica: Conflict and co-existence. *Development and Change* 33(1): 29–56. *Source*: http://people.duke.edu/~lcampbe/docs_lmc/Campbell_2002_Dev_Change.pdf (Accessed in November 2014).

Cao, S. and D.G.I. Kingston (2009). Biodiversity conservation and drug discovery: Can they be combined? The Suriname and Madagascar experiences. *Pharmaceutical Biology* 47(8): 809–823. Source *(Author manuscript)*: www.ncbi.nlm.nih.gov/pmc/articles/PMC2746688 (Accessed in May 2015).

Carley, M. and I. Christie (2000). *Managing Sustainable Development*. Earthscan Publications Ltd., London and Sterling, VA.

Carvalho, N.P. de (2003). *From the Shaman's Hut to the Patent Office: In Search of Effective Protection for Traditional Knowledge*. Washington University, St. Louis, Missouri, USA. *Source*: http://ls.wustl.edu/centeris/Confpapers (Accessed in January 2005).

Chander, R. (1996). A report on INMEDPLAN database. In: Tan, L.C., M.R. Pérez and M. Ibach (eds.), *Non-Timber Forest Product Databases*. CIFOR, Jakarta, Indonesia, pp. 53–58.

Chandiramani, N. (2002). Legal factors in TRIPS. *Economic and Political Weekly*, 19 January. *Source*: www.epw.org.in/showArticles.php (Accessed in February 2005).

Chandrasekharan, C. (1995). Terminology, definition and classification of forest products other than wood. In: Report of the International Expert Consultation on Non-Wood Forest Products. Non-Wood Forest Products – 3. FAO, Rome.

CIEL (2002). Preliminary Comments on Trade Related Issues in the 12 June Draft Plan of Implementation for the World Summit on Sustainable Development (WSSD). *Source*: www.ciel.org/Publications/ (Accessed in September 2012).

CIFOR (1999). *Who Counts Most? Assessing Human Well-Being in Sustainable Forest Management*. CIFOR, Jakarta, pp. 6–11. *Source*: www.cgiar.org/cifor (Accessed in September 2014).

CIPR (2002). Traditional knowledge and geographical indications (Chapter 4). In: *Integrating Intellectual Property Rights and Development Policy: Report of the Commission on Intellectual Property Rights*. Commission on Intellectual Property Rights, London, pp. 73–93. *Source*: www.iprcommission.org/graphic/documents/ (Accessed in October 2014).

Correa, C. (1999). Developing countries and the TRIPS agreement. An article (SUNS4497) from South-North Development Monitor (SUNS). *Source*: www.twn.my/title/correa-cn.htm (Accessed in May 2015).

Correa, C.M. (2001). Traditional Knowledge and Intellectual Property—A Discussion Paper. Quaker UN Office, Geneva. *Source*: www.geneva.quno.info/pdf/ (Accessed in August 2014).

Cullet, P. (1999). Intellectual property rights: For an alternative patents regime. *Frontline*, 9–22 October, Chennai, India. Vol. 16, Issue 21.

Dalton, R. (2004). Bioprospects less than golden. News feature. *Nature*, 10 June, 429: 598–600.

Dhar, B. and S. Chaturvedi (1998). *Implications of the Regime of Intellectual Property Protection for Biodiversity: A Developing Country Perspective*. RISNOPC, New Delhi. *As cited by* Ghate et al. (1999).

Driscoll, D.A., D.B. Lindenmayer, A.F. Bennett, M. Bode, R.A. Bradstock, G.J. Cary, et al. (2010). Fire management for biodiversity conservation: Key research questions and our capacity to answer them. *Biological Conservation* 143(9):1928–1939, doi: 10.1016/j.biocon.2010.05.026. *Source*: www.researchgate.net/profile/James_Watson8/publication/229314337…pdf (Accessed in May 2015).

Drissi, A., J. Girona, M. Cherki, G. Godàs, A. Derouiche, M. El Messal, R. Saile, A. Kettani, R. Solà, L. Masana and A. Adlouni (2004). Evidence of hypolipemiant and antioxidant properties of argan oil derived from the argan tree (*Argania spinosa*). *Clinical Nutrition* 23: 1159–1166. *Source*: www.researchgate.net/publication (Accessed in November 2014).

Dutfield, G. (1997). Between a rock and a hard place: Indigenous peoples, nation states and the multinationals. In: *Medicinal Plants for Forest Conservation and Health Care*. FAO Technical Papers, Non-Wood Forest Products-11. FAO, Rome, pp. 24–33.

———— (1999). The Public and Private Domains: Intellectual Property Rights in Traditional Ecological Knowledge. OIPRC (Oxford Intellectual Property Research Centre) Electronic Journal of Intellectual Property Rights, Working Paper 03/99. A revised version (March 2000) in *Science Communication* 21(2): 274–295. Source: www.researchgate.net/publication/240699081 (Accessed in December 2014).

———— (2000). *Intellectual Property Rights, Trade and Biodiversity: Seeds and Plant Varieties*. IUCN and Earthscan Publications Ltd., London. *As cited by* Gupta (2001).

Eberlee, J. (2000). Assessing the Benefits of Bioprospecting in Latin America. IDRC Reports. 21 January. *Source*: http://idl-bnc.idrc.ca/dspace/bitstream/10625/32433/1/114581.pdf (Accessed in November 2014).

Ekor, Martins (2013). The growing use of herbal medicines: issues relating to adverse reactions and challenges in monitoring safety. Published online 2014 January 10. *Frontiers in Pharmacology* 4: 177. *Source*: www.ncbi.nlm.nih.gov/pmc/articles/PMC3887317 (Accessed in April 2015).

EPW (2002). Biodiversity bill: A first step (Editorial). *Economic and Political Weekly*, December 28. Also available at: www.epw.in/editorials (Accessed in November 2014).

FAO (1995). Report of the International Expert Consultation on Non-Wood Forest Products. Yogyakarta, Indonesia, January. Technical Papers: Non-Wood Forest Products – 3. FAO, Rome.

FAO (1999). Towards a harmonized definition of non-wood forest products. *Unasylva* 50(3): 63–64. Also available at: www.fao.org/docrep/x2450e/x2450e0d.htm (Accessed in May 2015).

————— (2010). Global Forest Resources Assessment 2010 (Main Report). FAO Forestry Paper 163. Food and Agriculture Organisation of the United Nations, Rome.

————— (undated). How to enhance the participation of stakeholders in National Forest Programmes? *Source*: www.fao.org/forestry/foris/webview/forestry2/ (Accessed in July 2004).

Farnsworth, N.R. (1988). Screening plants for new medicines (Chapter 9). In Wilson, E.O. (ed.), *Biodiversity*. National Academy Press, Washington, DC. *Source*: http://www.ciesin.org/docs/002-256c/002-256c.html (Accessed in May 2015).

Farnsworth, N.R., O. Akerele, A.S. Bingel, D.D. Soejarto and Z. Guo (1985). Medicinal plants in therapy. *Bulletin of the World Health Organization* 63(6): 965–981. *Source*: www.ncbi.nlm.nih.gov/pmc/articles/PMC2536466/pdf/bullwho00089-0002.pdf (Accessed in March 2015).

FSI (2013). India State of Forest Report 2013. Forest Survey of India, Ministry of Environment and Forests, Government of India, Kaulagarh Road, Dehradun, India.

Gadgil, M. (2003). New roads, but a long way to go. *The Hindu*, New Delhi, 20 April. *Source*: www.thehindu.com/thehindu/mag/2003/04/20/stories/2003042000810100.htm (Accessed in October 2014).

Gamez, R. (2003). The Link between Biodiversity and Sustainable Development: Lessons from INBio's Bioprospecting Programme in Costa Rica. Paper presented in the conference on Biodiversity, Biotechnology and the Protection of Traditional Knowledge (4–6 April) at Washington University School of Law, St. Louis, Missouri (USA). *Source*: www.cbd.int/financial/bensharing/costarica-absinbio.pdf (Accessed in December 2014).

Gandhi, V.P. and N.T. Patel (2001). The impact of WTO on agricultural inputs. In: Datta, Samar K. and S.Y. Deodhar (coordinators), *Implications of WTO Agreements for Indian Agriculture*. CMA Monograph No. 191, Oxford & IBH Publishing Co. Pvt. Ltd., New Delhi/Calcutta, pp. 311–342.

Gangopadhyay, P.B. and D. Mohan (2005). Legislation on forest biodiversity conservation. In: Rawat, J.K., Shivendu K. Srivastava, Sas Biswas and H.B. Vasistha (eds.), *Conservation of Biodiversity in India: Status, Challenges and Efforts*. Indian Council of Forestry Research and Education, Dehradun, pp. 187–191.

Ganguli, P. (1998). *Gearing Up for Patents: The Indian Scenario*. University Press (India) Limited, Hyderabad.

Ghate, U., M. Gadgil and P.R. Sheshagiri Rao (1999). Intellectual property rights on biological resources: Benefiting from biodiversity and people's knowledge. *Current Science* 77(11): 1418–1425.

Gibbons, D. (2003). Green gold: The history and health care system of Costa Rica. Paper presented at '*Globalization and the Americas: An Undergraduate Conference on Scholarship and Career Paths*'. Michigan State University, October 23–24. *Source*: www.isp.msu.edu/clacs/conf-papers (Accessed in September 2004).

Girsberger, M.A., A.R. Kopše and F. Pythoud (1999). Draft guidelines on access and benefit sharing regarding the utilisation of genetic resources – Meeting of the expert panel on access and benefit sharing, San José, Costa Rica, 4–8 October. *Source*: www.cbd.int/doc/meetings/abs/absep-01/other/absep-01-guidelines-en.pdf (Accessed in November 2014).

Glaser, V. (2010). And another gold… Argan Oil—an emerging gold rush? Covalence Intern Analyst Papers. Covalence SA, Geneva. *Source*: www.cbd.int/financial/bensharing/morocco-arganoil.pdf (Accessed in July 2014).

Glowka, L. (1997). *A Guide to Designing Legal Frameworks to Determine Access to Genetic Resources*. IUCN, the World Conservation Union, Gland. *As cited by* Ghate, U., M. Gadgil and P.R. Sheshagiri Rao (1999).

Glowka, L., F. Burhenne-Guilmin, H. Synge, J. McNeely and L. Gundling (1994). *A Guide to the Convention on Biological Diversity*. Environmental Policy and Law Paper No. 3, The World Conservation Union (IUCN), Gland, Switzerland. *As cited by* Jermy (1995).

GOI (2000). Report of the Task Force on Conservation and Sustainable use of Medicinal Plants. Planning Commission, Government of India, New Delhi.

——— (2002). Keepers of the knowledge: Towards evolving a framework for benefit sharing (Chapter 6). In: Towards Sustainability: Stories from India. *WSSD Documents*, Johannesburg, 26 August–4 September, Document by Ministry of Environment and Forests, Government of India, pp. 20–23. *Source*: www.moef.nic.in/divisions/ic/wssd/doc3/chapter6/css/Chapter6.htm (Accessed in July 2014).

——— (2008). National Biodiversity Action Plan. Government of India, Ministry of Environment and Forests.

——— (2012). Guidelines for Processing of Patent Applications Relating to Traditional Knowledge and Biological Material. Intellectual Property India. Office of the Controller General of Patents, Designs and Trademarks. 18 December. *Source*: www.ipindia.nic.in/iponew/TK_Guidelines_18December2012.pdf (Accessed in June 2014).

GRAIN (2000). Global Trade and Biodiversity in Conflict. Series of exposés produced jointly by the Gaia Foundation and Genetic Resources Action International, Issue no. 4, pp. 20. *Source*: www.grain.org/article/entries/32-biodiversity-for-sale-dismantling-the-hype-about-benefit-sharing.pdf (Accessed in June 2014).

Green, E.C., K.J. Goodman and M. Hare (1999). Ethnobotany, IPR and Benefit-Sharing: The Forest People's Fund in Suriname. Indigenous Knowledge and Development Monitor, March. *Source*: https://app.iss.nl/ikdm/ikdm/ikdm/7-1/green.html (Accessed in May 2015)

Greene, S. (2004). Indigenous people incorporated? Culture as politics, culture as property in pharmaceutical bioprospecting. *Current Anthropology* 45(2): 211–237.

Guerin-McManus, M., K.C. Nnadozie and S.A. Laird (2002). Sharing financial benefits: Trust funds for biodiversity prospecting. In: Laird, S.A. (ed.), *Biodiversity and Traditional Knowledge: Equitable Partnerships in Practice*. Earthscan Publications Ltd., London (UK) and Sterling (USA).

Guerin-McManus, M., L.M. Famolare, I.A. Bowles, S.A.J. Malone, R.A. Mittermeier and A.B. Rosenfeld (1998). Bioprospecting in practice: A case study of the Suriname ICBG project and benefits sharing under CBD. Case study submitted to the Secretariat of the CBD. *Source*: www.icimod.org/resource/2252 (Accessed in October 2014).

Guillaume, D. and Z. Charrouf (2011). Argan oil: Monograph. *Alternative Medicine Review* 16(3): 275–279. *Source*: www.altmedrev.com/publications/16/3/275.pdf (Accessed in November 2014).

Gupta, A.K. (1999a). Conserving Biodiversity and Rewarding Associated Knowledge and Innovation Systems: Honey Bee Perspective. Paper presented at the First Commonwealth Science Forum–*Access, Bioprospecting, Intellectual Property Rights and Benefit Sharing and the Commonwealth*, Goa, 23–25 September. (A considerably revised and expanded version of the paper presented at World Trade Forum, Bern, August 27–29, 1999.) *Source*: www.sristi.org/papers/new/ (Accessed in January 2005).

——— (1999b). Environmental Implications of Intellectual Property Protection (IPP): Can individual and community conservation ethic and creativity be rewarded through IPP? Paper prepared for Brainstorming meeting on TRIPS and Environment, 11

October 1999, UNEP, Geneva. *Source*: www.iimahd.ernet.in/~anilg/morepub/index. html (Accessed in November 2004).

Gupta, A.K. (2001). How Can Asian Countries Protect Traditional Knowledge, Farmers Rights and Access to Genetic Resources through the Implementation or Review of the WTO TRIPS Agreement. Paper presented at the Joint ICTSD/CEE/HBF Regional Dialogue for Governments and Civil Society, organised by International Centre for Trade and Sustainable Development, Geneva at Chiang Mai, Thailand, March 29–30, 2001. *Source*: www.iimahd.ernet.in/~anilg/ (Accessed in October 2004).

——— (2002a). Role of intellectual property rights in the benefit-sharing arrangements surrounding the work of the Bio-resources Development and Conservation Programme. A case study based on the data collected from Nigeria. W.P.No.2002-08-03. *Source*: www. iimahd.ernet.in; www.sristi.org/papers/new/ (Accessed in February 2005). *Also published as*: WIPO-UNEP Study on the Role of Intellectual Property Rights in the Sharing of Benefits Arising from the Use of Biological Resources and Associated Traditional Knowledge. *Source*: www.wipo.int/tk/en/publications/ (Accessed in February 2005).

——— (2002b). Value addition to local Kani tribal knowledge: patenting, licensing and benefit-sharing. Working Paper No. 2002-08-02. Indian Institute of Management, Ahmedabad. *Source*: www.iimahd.ernet.in. (Accessed in February 2005). *Also in*: WIPO-UNEP Study on the Role of Intellectual Property Rights in the Sharing of Benefits Arising from the Use of Biological Resources and Associated Traditional Knowledge, pp. 103-123. *Source*: www.wipo.int/tk/en/publications/ (Accessed in February 2005).

——— (2003). Rewarding Conservation of Biological and Genetic Resources and Associated Traditional Knowledge and Contemporary Grassroots Creativity. Working Paper No. 2003-01-06. Indian Institute of Management, Ahmedabad. *Source*: www. sristi.org/papers/new/ (Accessed in February 2005).

——— (2005). CBD and TRIPS: Empowering knowledge rich, economically poor people through IPR reforms. Paper presented at the National Seminar on TRIPS–CBD and Subsidy Issues at the WTO on 25 August 2005, New Delhi. *Source*: www.iimahd.ernet. in/~anilg/selectedpub.php (Accessed in December 2012).

Hall, K. (2003). Background Paper: What is Bioprospecting and what are our International Commitments? University of Auckland, School of Geography and Environmental Science, New Zealand. *Source*: www.med.govt.nz/ers/nat-res/bioprospecting/general-information (Accessed in September 2004).

Halle, M. and N. Borregaard (2004). The Trade and Environment Agenda Post-Johannesburg. In: Bigg, Tom (ed.), *Survival for a Small Planet: The Sustainable Development Agenda*. International Institute for Environment and Development/Earthscan, London, pp. 32–45.

Hurlbut, David (1994). Fixing the biodiversity convention: Toward a special protocol for related intellectual property. *Natural Resources Journal* 34(2): 379–409.

ICFRE (2000). National Forestry Research Plan. Indian Council of Forestry Research and Education, New Forest, Dehradun.

ICIMOD (2010). Towards an Access and Benefit Sharing Framework Agreement for the Genetic Resources and Traditional Knowledge of the Hindu Kush-Himalayan Region. *Draft for discussion*. ICIMOD, Kathmandu, Nepal. *Source*: www.icimod.org (Accessed in August 2014).

Iqbal, M. (1993). International trade in non wood forest products: An overview. Working Paper No. Misc/93/11, FAO, Rome. *Source*: www.fao.org/docrep (Accessed in July 2014).

——— (1995). Trade restrictions affecting international trade in non-wood forest products. FAO Technical Papers, Non-Wood Forest Products-8. FAO, Rome.

Janick, J. and R.E. Paull (eds.) (2008). *The Encyclopedia of Fruit and Nuts*, pp. 822–823. CAB International, Wallingford, UK.

Jermy, C. (1995). Legal and ethical aspects of biodiversity assessment: The convention on biological diversity. In: Jermy, C., D. Long, M. Sands, N. Stork and S. Winser (eds.), *Biodiversity Assessment: A Guide to Good Practice*. Department of the Environment/ HMSO, London, pp. 45–69.

Johnson, M. (1992). Research on traditional environmental knowledge: Its development and its role. In: Johnson, M. (ed.), *Lore: Capturing Traditional Environmental Knowledge*. IDRC, Ottawa, Canada, pp. 3–22. *Source:* www.idrc.ca/EN/Resources/Publications/ openebooks/644-6/index.html (Accessed in May 2014).

Jones, E.T., R.J. McLain and J. Weigand (2002). *Non-timber forest products in the United States*. University Press of Kansas, Lawrence, USA. *As cited in* the FAO document—Case Study No. 7: Impact of Cultivation and Gathering of Medicinal Plants on Biodiversity: Global Trends and Issues, authored by Schippmann, U., A.B. Cunningham and D.J. Leaman. *Source*: www.fao.org/3/a-aa010e/AA010E00.pdf (Accessed in January 2014).

Kaimowitz, D. and D. Sheil (2010). Conserving what and for whom? Why conservation should help meet basic needs in the tropics (Chapter 23). In: Roe, D. and J. Elliott (eds.), *The Earthscan Reader in Poverty and Biodiversity Conservation*. Earthscan, London and Washington DC, pp. 245–259.

Kamboj, V.P. (2000). Herbal medicine. *Current Science* 78(1): 35–39. *Source*: www.iisc.ernet. in/~currsci/jan102000/GENERALARTICLES.pdf (Accessed in April 2015).

Kameswari, V.L.V. (2002). Collective management of forest resources in India: Lessons for policy formulation and stakeholders involvement. Paper presented at 'The Commons in an Age of Globalisation,' the Ninth Conference of the International Association for the Study of Common Property, Victoria Falls, Zimbabwe, June 17–21, 2002. *Source*: http:// dlc.dlib.indiana.edu/archive/ (Accessed in February 2005).

Karki, M. (2003). Organic conversion and certification: A strategy for improved value-addition and marketing of medicinal plants products in the Himalayas. *Indian Forester* 129(1): 130–142.

Kate, K. ten and A. Wells (2001). Preparing a national strategy on access to genetic resources and benefit-sharing: A pilot study. The study supported financially by the UK Department of the Environment, Transport and the Regions and the UNDP/UNEP Biodiversity Planning Support Programme; funded by the Global Environment Facility. *Source*: http://teebforbusiness.earthmind.net/files/Preparing_a_National_Strategy_on_ Access_to_Genetic_Resources_and_Benefit-Sharing.pdf (Accessed in December 2014).

Kaul, O.N. (2005). Conserving biodiversity through participatory management (Chapter 23). In: Rawat, J.K., Shivendu K. Srivastava, Sas Biswas and H.B. Vasistha (eds.), *Conservation of Biodiversity in India: Status, Challenges and Efforts*. Indian Council of Forestry Research and Education, Dehradun, pp. 166–169.

Khalil, M. (1995). Biodiversity and the conservation of medicinal plants: Issues from the perspective of the developing world. In: Swanson, T.M. (ed.), *Intellectual Property Rights and Biodiversity Conservation: An Interdisciplinary Analysis of the Values of Medicinal Plants*. Cambridge University Press, Cambridge, USA, pp. 232–253.

Khoshoo, T.N. (1995). Biodiversity, bioproductivity and biotechnology. *Ambio* 24(4): 251–253.

Kiene, T. (2011). *The Legal Protection of Traditional Knowledge in the Pharmaceutical Field: An Intercultural Problem on the International Agenda*. Waxmann Verlag, Germany.

Kothari, A. and R.V. Anuradha (1997). Biodiversity, intellectual property rights, and the GATT agreement: How to address the conflicts? *Biopolicy* (online journal) 2, 4. *Source*: http://bioline.utsc.utoronto.ca/archive (Accessed in January 2005).

Koziell, I. and C. McNeill (2004). Poverty reduction through conservation and sustainable use of biodiversity. In: Bigg, Tom (ed.), *Survival for a Small Planet: The Sustainable Development Agenda*. International Institute for Environment and Development/ Earthscan, London, pp. 247–257.

Kuipers, S.E. (1997). Trade in medicinal plants. In: Bodeker, G., K.K.S. Bhat, Jeffrey Burly and Paul Vantomme (eds.), *Medicinal Plants for Forest Conservation and Health Care'* Technical Papers: Non-Wood Forest Products-11. FAO, Rome, pp. 45–59.

Kumar, N., N.C. Saxena, Y. Alagh and A.K. Mitra (2000). *India: Alleviating Poverty through Forest Development*. World Bank, Washington. *As cited by* Saxena, N.C. (2003).

Kumaraswamy, S. and K. Kunte (2013). Integrating biodiversity and conservation with modern agricultural landscapes. *Biodiversity and Conservation* 22: 2735–2750.

L'Oréal Canada. (undated). Responsible sourcing of argan oil: A case study. *Source*: www. businessbiodiversity.ca/documents/8-LOreal.pdf (Accessed in September 2013).

Laird, S.A. and E.E. Lisinge (2002). Protected area research policies: Developing a basis for equity and accountability. In: Laird, S.A. (ed.), *Biodiversity and Traditional Knowledge: Equitable Partnerships in Practice*. Earthscan Publications Ltd., London and Sterling (USA), pp. 161–164.

Laird, S.A. and K. ten Kate (1999). The Commercial Use of Biodiversity: Access to Genetic Resources and Benefit-Sharing. Report prepared for the European Commission by Royal Botanic Gardens, Kew (UK). Earthscan, London.

Laird, S.A. and K. ten Kate (2002). Biodiversity prospecting: The commercial use of genetic resources and best practice in benefit-sharing. In: Laird, S.A. (ed.), *Biodiversity and Traditional Knowledge: Equitable Partnerships in Practice*. Earthscan Publications Ltd., London and Sterling (USA).

Laird, S.A. and R. Wynberg (2002). Institutional policies for biodiversity research (Chapter 3). In: Laird, Sarah A. (ed.), *Biodiversity and Traditional Knowledge: Equitable Partnership in Practice*. Earthscan Publications, London, Sterling VA (USA), pp. 39–76.

Lapham, N.P. and R.J. Livermore (2010). Striking a Balance: Ensuring Conservation's Place on the International Biodiversity Assistance Agenda (Chapter 8). In: Roe, D. and J. Elliott (eds.), *The Earthscan Reader in Poverty and Biodiversity Conservation*. Earthscan, London and Washington, DC, pp. 78–103.

Lash, J. (1993). Foreword. In: Reid, W.V., S.A. Laird, C.A. Meyer, R. Gamez, A. Sittenfeld, D.H. Janzen, M.A. Gollin and C. Juma (eds.), *Biodiversity Prospecting: Using Genetic Resources for Sustainable Development*. World Resources Institute, USA.

Lawrence, A. (2003). The unmeasurable whole: Assessing forest biodiversity with multiple stakeholders. Paper submitted to the XII World Forestry Congress, Quebec. *Source*: http://r4d.dfid.gov.uk/PDF/Outputs/Forestry/R7475_-_Assessing_Forest_Biodiversity. pdf (Accessed in February 2014).

Lesser, W. (1998). *Sustainable Use of Genetic Resources under the Convention on Biological Diversity: Exploring Access and Benefit Sharing Issues*. CAB International, New York.

Lewington, A. (1993). *Medicinal Plant and Plant Extracts: A Review of Their Importation into Europe*. Traffic International, Cambridge, UK.

Lybbert, T.J., A. Aboudrare, D. Chaloud, N. Magnan and M. Nash (2011). Booming markets for Moroccan argan oil appear to benefit some rural households while threatening the endemic argan forest. *Proceedings of the National Academy of Sciences* 108(34): 13963–13968. *Source*: http://agecon.ucdavis.edu/people/faculty/travis-lybbert/publications (Accessed in May 2014).

Lybbert, T.J., N. Magnan and A. Aboudrare (2010). Household and local forest impacts of Morocco's argan oil bonanza. *Environment and Development Economics* 15: 439–464. *Source*: http://agecon.ucdavis.edu/people/faculty/travis-lybbert/publications (Accessed in May 2014).

Maudgal, S. and M. Kakkar (1996). Biodiversity and national issues. In: Gujral, G.S. and V. Sharma (eds.), *Changing Perspectives of Biodiversity Status in the Himalaya*. British Council, New Delhi, pp. 181–186.

Maxwell, D. (1998). Can qualitative and quantitative methods serve complementary purposes for policy research? Evidence from Accra. FCND Discussion Paper No. 40. Food Consumption and Nutrition Division, International Food Policy Research Institute, Washington DC. *Source*: www.ifpri.org/sites/default/files/publications/dp40.pdf (Accessed in July 2014).

McLaughlin, R.J. (2003). Foreign access to shared marine genetic materials: Management options for a Quasi-Fugacious resource. *Ocean Development & International Law* 34: 297–348.

McManis, C.R. (2003). Intellectual property, genetic resources and traditional knowledge protection: Thinking globally, acting locally. *Cardozo Journal of International and Comparative Law* 11(2): 547–583.

———— (2004). Fitting traditional knowledge protection and biopiracy claims into the existing intellectual property and unfair competition framework (Chapter 12). In: Ong, Burton (ed.), *Intellectual Property and Biological Resources*. Marshall Cavendish Academic, Singapore, pp. 425–510.

Medaglia, J.C., F. Perron-Welch and O. Rukundo (2012). Overview of National and Regional Measures to Access on Genetic Resources and Benefit-Sharing: Challenges and Opportunities in Implementing the Nagoya Protocol. Centre for International Sustainable Development Law. *Source*: www.cisdl.org/biodiversity…/CISDL_Overview_of_ABS_Measures_2nd_Ed.pdf. (Accessed in May 2014).

Meena, A.K., P. Bansal and S. Kumar (2009). Plants-herbal wealth as a potential source of ayurvedic drugs. *Asian Journal of Traditional Medicines* 4(4): 152–168. *Source*: www.asian-jtm.com/qikan/manage/wenzhang/AJTM2009-4(4)-5.pdf (Accessed in April 2015).

Mehrotra, S. (2003). Standardization and quality control—The mandatory requirement of herbal drug industries. *Indian Forester* 129(2): 233–242.

Mehta, P.S. (2002). *WTO and India: An Agenda for Action in Post Doha Scenario*. CUTS Centre for International Trade, Economics & Environment, Jaipur (India).

Menon, V. (1996). Impact of trade on Himalayan biodiversity. In: Gujral, G.S. and V. Sharma (eds.), *Changing Perspectives of Biodiversity Status in the Himalaya*. British Council, New Delhi, pp. 139–147.

Millennium Ecosystem Assessment (2005). Beattie, Andrew J. (Coordinating Lead Author), Wilhelm Barthlott, Elaine Elisabetsky, Roberta Farrel, Chua Teck Kheng, Iain Prance (Lead Authors), Joshua Rosenthal, David Simpson, Roger Leakey, Maureen Wolfson, Kerry ten Kate (Contributing Authors) and Sarah Laird (Review Editor). New Products and Industries from Biodiversity. In: Hassan, Rashid, Robert Scholes and Neville Ash (eds.), *Ecosystem and Human Well-being: Current State and Trends* (Volume 1), pp. 271–295. *Source*: www.maweb.org/documents/document.279.aspx.pdf (Accessed in May 2014).

Miller, J.S. (2003). Impact of the Convention on Biological Diversity: The Lessons of Ten Years of Experience with Models for Equitable Sharing of Benefits. Document posted in August 2003. *Source*: http://law.wustl.edu/centeris/Confpapers/PDFWrdDoc/ (Accessed in January 2005).

Miller, M.A.L. (1995). *The Third World in Global Environmental Politics.* Open University Press, Buckingham.

Mora, M.A. (1996). Biodiversity information management system. In: Tan, L.C., M.R. Pérez and M. Ibach (eds.), *Non-Timber Forest Product Databases.* Special Publication. Centre for International Forestry Research (CIFOR), Jakarta, Indonesia, pp. 37–43. *Source*: www.cifor.cgiar.org/publications/ (Accessed in September 2004).

Moran, K. (1999). Health: Indigenous Knowledge, Equitable Benefits. IK Notes No. 15. World Bank. *Source*: http://documents.worldbank.org/curated/en/1999/12/1671225/ health-indigenous-knowledge-equitable-benefits (Accessed in February 2014).

———— (2000). Bioprospecting: Lessons from benefit-sharing experiences. *International Journal of Biotechnology* 2(1/2/3): 132–144.

Morgera, E., E. Tsioumani and M. Buck (2014). Unraveling the Nagoya Protocol: A Commentary on the Nagoya Protocol on Access and Benefit-Sharing to the Convention on Biological Diversity. Introduction. Brill/Martinus Nijhoff.

Mott, R. (1993). The GEF and the conventions on climate change and biological diversity. *International Environmental Affairs* 5(4): 299–312.

Moussouris, Y. and A. Pierce (2000). Biodiversity links to cultural identity in southwest Morocco: The situation, the problems and proposed solutions. *Aridlands Newsletter.* No. 48, November/December. *Source*: http://ag.arizona.edu/oals/ALN/aln48toc.html (Accessed in September 2013).

Mulder, M.B. and P. Coppolillo (2005). *Conservation: Linking Ecology, Economics, and Culture.* Princeton University Press, New Jersey (USA).

Murad, W., A. Azizullah, M. Adnan, A. Tariq, K.U. Khan, S. Waheed and A. Ahmad (2013). Ethnobotanical assessment of plant resources of Banda Daud Shah, District Karak, Pakistan. *Journal of Ethnobiology and Ethnomedicine* 9(77). Published: 22 November. *Source*: www.ethnobiomed.com/content/9/1/77 (Accessed in October 2014).

Myers, N. (1995). Population and biodiversity. *Ambio* 24(1): 56–57.

Nair, M.D. (2000). Winning the war against bio—Colonisation. *The Hindu*, New Delhi, 17 May.

———— (2001). Protection of trade secrets, undisclosed information. *The Hindu*, New Delhi, 22 November.

———— (2002a). Issues on TRIPS of Concern to India. *The Hindu*, New Delhi, 23 September.

———— (2002b). Legislative options for protection. *The Hindu*, New Delhi, 22 April.

NALSAR (undated). Agreement on Trade-Related Aspects of Intellectual Property Rights. Recent Articles. *Source*: www.nalsarpro.org/PL/Articles/TRIPSAgreement.pdf (Accessed in November 2014).

NBA (2013a). *Guidelines for Operationalisation of Biodiversity Management Committees.* National Biodiversity Authority, Chennai.

———— (2013b). *Report of the 7th National Meeting of State Biodiversity Boards held on 27–28 January in Chennai.* National Biodiversity Authority, Chennai.

———— (undated). *Guidelines for Identification, Notification and Management of Biodiversity Heritage Sites.* National Biodiversity Authority, Chennai, India.

NIH (undated). *Drug Discovery and Biodiversity among the Maya of Mexico.* Research Portfolio Online Reporting Tools, NIH, USA. *Source*: http://projectreporter.nih.gov/pr_Prj_info_ desc_dtls.cfm?projectnumber=5U01TW001009-03&print=yes (Accessed in May 2015). Also available from: www.labome.org/grant/u01/tw/drug/discovery/drug-discovery-and-biodiversity-among-the-maya-of-mexico-6188573.html (Accessed in May 2015).

Nnadozie, K.C., M.M. Iwu, E.N. Sokomba and C. Obialor (2002). Case Study 11.2: Nigeria's Fund for Integrated Rural Development and Traditional Medicine (FIRD-TM). In: Laird, Sarah A. (ed.), *Biodiversity and Traditional Knowledge: Equitable Partnership in Practice*. Earthscan, pp. 340–342. *Source*: books.google.co.in/books?isbn=1136534601 (Accessed in May 2014).

NNBP (1999). Thematic paper to support the input for the Traditional Knowledge Focal Group to the National Biodiversity Strategy and Action Plan. Draft by Uariua-Kakujaha, K, M. Katjiua and A. Mosimane. Namibian National Biodiversity Programme, Traditional Knowledge Sector Paper, Windhoek, 28pp.

Nygren, A. (1998). Environment as discourse: Searching for sustainable development in Costa Rica. *Environmental Values* 7: 201–222.

Park, C. (2010). Implementation of India's Patent Law: A review of patents granted by the Indian Patent Office. In: Chaudhuri, Sudip, Chan Park and K.M. Gopakumar (eds.), *Five Years into the Product Patent Regime: India's Response*, pp. 73–104. UNDP, New York. *Source*: www.undp.org, also available at: apps.who.int/medicinedocs/documents/s17761en/s17761en.pdf (Accessed in May 2014).

Patnaik, J.K. (1997). *India and the GATT: Origin, Growth and Development*. APH Publishing Corporation, New Delhi.

Prashantkumar, P. and G.M. Vidyasagar (2008). Traditional knowledge on medicinal plants used for the treatment of skin diseases in Bidar district, Karnataka. *Indian Journal of Traditional Knowledge* 7(2):273–276. *Source*: www.niscair.res.in/Sciencecommunication/ResearchJournals/rejour/ijtk/Fulltextsearch/2008/20April2008/IJTK (Accessed in April 2015).

Raghavan, C. (2001). Shaky start for WIPO meet on IPRs, Genetic Resources. Report from Geneva, 3 May. *Source*: www.twn.my/title/shaky.htm (Accessed in January 2015).

Raja, K. (2001). WIPO meet next week on IPRs and traditional knowledge. Report from Geneva, 25 April. SUNS document no. SUNS4884. *Source*: www.twn.my/title/wipo.htm (Accessed in January 2015).

Ramakrishnappa, K. (undated). Impact of Cultivation and Gathering of Medicinal Plants on Biodiversity: Case Studies from India. FAO Document: Case Study No. 8. *Source*: www.fao.org/docrep/005/y4586e/y4586e09.htm (Accessed in April 2014).

Ramanna, A. (2002). Policy implications of India's patent reforms: Patent applications in the post-1995 era. *Economic and Political Weekly*, 25 May.

Rao, M.B. (2001). *WTO and International Trade*. Vikas Publishing House, New Delhi.

Rau, B.S., G.G. Nair and P.V. Appaji (2012). Current status of pharmaceutical patenting in India. *Pharma Times* 44(7): 13–15. *Source*: www.pharmexcil.org/uploadfile/ufiles/5CurrentStatusPharmaPatentingInIndia01july2012.pdf (Accessed in May 2014).

Reid, W.V. (1993). Bioprospecting: A force for sustainable development. *Environmental Science and Technology* 27(9): 1730–1732.

RIS (2003). World Trade and Development Report 2003: Cancun and Beyond. Research and Information System for the Non-Aligned and Other Developing Countries (RIS), New Delhi and Academic Foundation, New Delhi.

Robinson, D. and E. Defrenne (2009). Argan: A Case Study on ABS. Union for Ethical Bio Trade (UEBT). *Source*: www.ethicalbiotrade.org/dl/benefit-sharing/UEBT_Argan/D_ROBINSON_AND_E_DEFRENNE_final.pdf (Accessed in May 2014).

Rosenthal, J.P. (1997a). Equitable Sharing of Biodiversity Benefits: Agreements on Genetic Resources. In: Investing in Biological Diversity: Proceedings of the Cairns Conference, OECD. Paper presented at the International Conference on Incentive Measures for the Conservation and the Sustainable Use of Biological Diversity, Cairns, Australia, 25–28 March 1996. *Source*: http://citeseerx.ist.psu.edu/viewdoc/download (Accessed in June 2014).

Rosenthal, J.P. (1997b). Integrating Drug Discovery, Biodiversity Conservation, and Economic Development: Early Lessons from the International Cooperative Biodiversity Groups (Chapter 13). In: Grifo, Francesca and Joshua Rosenthal (eds.), *Early Lessons for the International Cooperative Biodiversity Groups: Biodiversity and Human Health*. Island Press, Washington DC. *Source*: www.icbg.org/sphider/search.php (Accessed in May 2014).

———— (1998). The ICBG Programme: A Benefit-Sharing Case Study for the Conference of Parties to Convention on Biological Diversity, submitted in 1998. *Source*: www.cbd.int/financial/bensharing/unitedstates-icbg.pdf (Accessed in November 2014; previously accessed in January 2005 from: www.biodiv.org/doc/meetings/cop).

———— (2006). Politics, culture, and governance in the development of prior informed consent in indigenous communities. *Current Anthropology* 47(1): 119–142. *Source*: www.icbg.org/documents/PIC_and_Indigenous_pop_CA.pdf (Accessed in May 2014).

Ruiz, M. (2002). The International Debate on Traditional Knowledge as *prior art* in the Patent System: Issues and Options for Developing Countries. Centre for International Environmental Law. *Source*: www.ciel.org/Publications/PriorArt_ManuelRuiz_Oct02.pdf (Accessed in August 2013).

Sahai, S. (2003). An IPR agenda for India at Cancun. *Economic Times*, New Delhi, 1 September.

Sarin, Y.K. (2003a). Medicinal plant raw materials for Indian drug and pharmaceutical industry: I—An appraisal of resources. *Indian Forester* 129(1): 3–24.

———— (2003b). Medicinal plant raw materials for Indian drug and pharmaceutical industry: II—Problems and prospects of development of resources. *Indian Forester* 129(2): 143–153.

Saxena, N.C. (2003). From monopoly to de-regulation of NTFPs: Policy shifts in Orissa (India). *International Forestry Review* 5(2): 168–176.

Sayer, J., C. Elliott and S. Maginnis (2005). Protect, manage and restore: Conserving forests in multifunctional landscapes (Chapter 21). In: Sayer, Jeffrey (ed.), *The Earthscan Reader in Forestry and Development*, pp. 415–421. Earthscan, London and Sterling VA (USA).

SCBD (1998a). Synthesis of Case-Studies on Benefit Sharing. Information Document UNEP/CBD/COP/4/INF/7, Fourth Meeting of the COP to the CBD, Bratislava, 4–15 May. *Source*: www.cbd.int/doc/meetings/cop/cop-04/information/cop-04-inf-07-en.pdf (Accessed in October 2013).

———— (1998b). Report of the International Workshop on 'Best Practices' for Access to Genetic Resources, Cordoba, 16–17 January. (Document code: cop-04-info-10). Submission by the European Commission and the Government of Germany in the Fourth Meeting of COP to the CBD, Bratislava, Slovakia, 4-15 May. Secretariat of the Convention on Biological Diversity. *Source*: www.cbd.int/doc/meetings/cop/cop-04/information/cop-04-inf-10-en.pdf (Accessed in October 2013).

———— (2002). Bonn Guidelines on Access to Genetic Resources and Fair and Equitable Sharing of the Benefits Arising out of their Utilization. Secretariat of the Convention on Biological Diversity, Montreal. *Source*: www.cbd.int/abs/bonn/ (Accessed in September 2014).

———— (2007). Development of Elements of *sui generis* Systems for the Protection of Traditional Knowledge, Innovations and Practices to Identify Priority Elements. Note by the Executive Secretary for *ad hoc* open-ended inter-sessional Working Group on Article 8(j) and related provisions of the CBD Fifth Meeting, Montreal, 15–19 October. Document Code: UNEP/CBD/WG8J/5/6, September 20. *Source*: www.cbd.int/doc/meetings/tk/wg8j-05/official/wg8j-05-06-en.pdf (Accessed in November 2014).

Secrett, C. (2004). The politics of radical partnerships: Sustainable development, rights and responsibilities. In: Bigg, Tom (ed.) *Survival for a Small Planet: The Sustainable Development Agenda*. International Institute for Environment and Development/Earthscan, London, pp. 152–172.

Shankar, D. and B. Majumdar (1997). Beyond the biodiversity convention: The challenges facing the biocultural heritage of India's medicinal plants. In: Bodeker, G., K.K.S. Bhat, Jeffrey Burly and P. Vantomme (eds.), *Medicinal Plants for Forest Conservation and Health Care*. FAO Technical Papers, Non-Wood Forest Products-11. FAO, Rome, pp. 87–99.

Sharma, D. (1996). Biodiversity and international issues. In: Gujral, G.S. and V. Sharma (eds.), *Changing Perspectives of Biodiversity Status in the Himalaya*. British Council, New Delhi, pp. 173–180.

——— (2000). Patently wrong on medicinal plants. *Invention Intelligence*, November–December, pp. 253–255.

——— (2004). Selling Biodiversity: Benefit sharing is a dead concept. *Source*: www.mindfully.org/WTO/ (Accessed in February 2014).

Sharma, S.C. and M. Negi (2005). Biodiversity conservation imbroglio: Theoretical issues (Chapter 2). In: Rawat, J.K., Shivendu K. Srivastava, Sas Biswas and H.B. Vasistha (eds.), *Conservation of Biodiversity in India: Status, Challenges and Efforts*. Indian Council of Forestry Research and Education, Dehradun, pp. 8–15.

Sharma, V.P., S.K. Datta and P. Sharma (2001). Trade in non-timber forest products in India: Some issues and concerns. In: Datta, Samar K. and S.Y. Deodhar (coordinators), *Implications of WTO Agreements for Indian Agriculture*. CMA Monograph No. 191, Oxford & IBH Publishing Co. Pvt. Ltd., New Delhi/Calcutta, pp. 540–551.

Sheldon, J.W. and M.J. Balick (1995). Ethnobotany and the search for balance between use and conservation. In: Swanson, T.M. (ed.) *Intellectual Property Rights and Biodiversity Conservation: An Interdisciplinary Analysis of the Values of Medicinal Plants*. Cambridge University Press, Cambridge, USA, pp. 45–64.

Shiva, M.P. (1998). *Inventory of Forest Resources for Sustainable Management and Biodiversity Conservation*. Centre for Minor Forest Products, Dehradun.

Shiva, V. (2000). *Stolen Harvest: The Hijacking of the Global Food Supply*. India Research Press, New Delhi.

——— (2001). *Patents: Myths and Reality*. Penguin Books India, New Delhi.

Simpson, D. and R. Sedjo (2004). Golden rule of economics yet to strike prospectors. Correspondence. *Nature*, 12 August, 430: 723.

Singh, A.G., A. Kumar and D.D. Tewari (2012). An ethnobotanical survey of medicinal plants used in Terai forest of western Nepal. *Journal of Ethnobiology and Ethnomedicine* 8(19). Published online May 16. *Source*: www.ncbi.nlm.nih.gov/pmc/articles/PMC3473258 (Accessed in October 2014).

Singh, S. (1999). Traditional Knowledge under Commercial Blanket. South-North Development Monitor (SUNS) No. 4545, 5 November. *Source*: www.grain.org/article/entries/1958-indigenous-raise-debate-in-geneva (Accessed in May 2014).

Singh, S., V. Sankaran, H. Mander and S. Worah (2000). Strengthening Conservation Cultures: Local Communities and Biodiversity Conservation. Man and the Biosphere Programme, UNESCO, Paris.

SUNS (1999). Indigenous People Criticise WIPO Approach. Report by Martin Khor. *South-North Development Monitor*, Third World Network, Geneva. *Source*: www.gene.ch/info4action/1999/Nov/msg00031.html (Accessed in January 2015).

Suzman, J. (2001). An introduction to the Regional Assessment of the Status of the San in Southern Africa. Regional Assessment of the Status of the San in Southern Africa Report Series. Report No. 1 of 5. Legal Assistance Centre, Windhoek. *Source*: www.lac.org.na/projects/lead/Pdf/sanintro.pdf (Accessed in May 2014).

Swanson, T.M. (1995). The appropriation of evolution values: an institutional analysis of intellectual property regimes and biodiversity conservation. In: Swanson, T.M. (ed.) *Intellectual Property Rights and Biodiversity Conservation: An Interdisciplinary Analysis of the Values of Medicinal Plants*. Cambridge University Press, Cambridge, USA, pp. 141–175.

Swiderska, K. (undated). Sharing the benefits from genetic resource use. Biodiversity in Development (Biodiversity Brief 3). IUCN, pp. 4. *Source*: www.iucn.org/themes/ (Accessed in October 2004).

Taleb, M.S. (2014). Argan (*Argania spinosa* (L.) Skeels) in Morocco: Function, management, and Access and Benefit Sharing. Poster (No. WCA2014-257) presented at World Congress on Agroforestry, New Delhi, 10–14 February. Institut Scientifique, Université Mohammed V-Agdal, Rabat, Morocco. *Source*: http://blog.worldagroforestry.org/index.php/2014/03/24/moroccos-tree-of-life-in-decline/#sthash.OoMeBJkb.dpuf (Accessed in May 2014).

Tan, L.C., M.R. Pérez and M. Ibach (1996). Survey of non-timber forest product databases. In: Tan, L.C., M.R. Pérez and M. Ibach (eds.), *Non-Timber Forest Product Databases*. CIFOR, Jakarta, Indonesia, pp. 1–21.

Tarasofsky, R.G. (2002). *Towards a mutually supportive relationship between the Convention on Biological Diversity and the World Trade Organization: An Action Guide*. IUCN, Gland, Switzerland and Cambridge, UK. v + 26pp. *Source*: https://portals.iucn.org/library/efiles/documents/2002-003.pdf (Accessed in July 2014).

Taylor, D.A. (1999). Requisites for thriving rural non-wood forest product enterprises. *Unasylva* 50(198): 3–8.

TED (undated). Merck-INBIO Plant Agreement (MERCK). Trade and Environment Database (TED). *Source*: www.american.edu/projects/mandala/TED/ (Accessed in October 2004). Also available: www1.american.edu/ted/MERCK.HTM (Accessed in April 2015).

Tellez, V.M. (undated). Recognising the traditional knowledge of the San people: The Hoodia case of benefit-sharing. *Source*: www.ipngos.org/.../Hoodia_case_of_benefit_sharing.pdf (Accessed in May 2014).

The Hindu (2004). Move to patent tribal knowledge. Report by T. Nandakumar. *The Hindu*, New Delhi, 20 January.

The Lancet (1994). Pharmaceuticals from plants: Great potential, few funds. *The Lancet* 343: 1513–1515.

The Observer (2001). In Africa the Hoodia Cactus Keeps Men Alive: Now its secret is stolen to make us thin. 17 June 2001. *Source*: www.theguardian.com/world/2001/jun/17/internationaleducationnews. businessofresearch (Accessed in May 2014).

Thorpe, P. (2002). Study on the Implementation of the TRIPS Agreement by Developing Countries. Commission on Intellectual Property Rights – Study Paper 7. *Source*: www.iprcommission.org/papers/pdfs/study_papers/sp7_thorpe_study.pdf (Accessed in December 2013).

TIFAC (2000). Intellectual Property Rights Bulletin, a bulletin from Technology Information, Forecasting and Assessment Council (TIFAC), Department of Science and Technology, GOI, New Delhi, Vol. 6 No. 9, pp. 1–3.

TWN (2013). Costa Rica's INBio, nearing collapse, surrenders its biodiversity collections and seeks government bailout. Article by Edward Hammond, 20 April. *Source:* www.twn.my/title2/biotk/2013/biotk130401.htm (Accessed in May 2015).

——— (2015). Amid controversy and irony, Costa Rica's INBio surrenders biodiversity collections and lands to the State. Report by Edward Hammond, 2 April 2015. *Source*: www.twn.my/title2/biotk/2015/btk150401.htm (Accessed in May 2015).

UNCTAD (2000). Benefit Sharing: Experience of Costa Rica. Paper prepared by the Government of Costa Rica for the Second Regional Workshop of the UNCTAD Project on 'Strengthening Research and Policy Making Capacity on Trade and Environment in the Developing Countries.' 30 May–2 June, Havana, Cuba. *Source*: http://r0.unctad.org/trade_env/discpapers2.htm (Accessed in September 2014).

UNDP (2012). Recognition of Community Rights under Forest Rights Act in Madhya Pradesh and Chhattisgarh: Challenges and Way Forward. Final Report, July 2011. Prepared by SAMARTHAN – Centre for Development Support. *Source*: www.undp.org/content/dam/india/docs/DG/recognition-of-community-rights-under-forest-rights-act-in-madhya-pradesh-and-chhattisgarh-challenges-and-way-forward.pdf (Accessed in November 2014).

USDA (undated). Conservation of Biological Diversity. Research Topics. Pacific South West Research Station, Berkeley, US Forest Service, US Department of Agriculture. *Source*: www.fs.fed.us/psw/topics/ecosystem_processes/sierra/bio_diversity (Accessed on 7 May 2015).

Utkarsh, G., M. Gadgil, S. Dasgupta and A. Chhatre (2005). Protecting people's knowledge in the regime of intellectual property rights. In: Rawat, J.K., Shivendu K. Srivastava, Sas Biswas and H.B. Vasistha (eds.), *Conservation of Biodiversity in India: Status, Challenges and Efforts*. Indian Council of Forestry Research and Education, Dehradun, pp. 116–126.

Vantomme, P. (2001). Production and trade opportunities for Non-Wood Forest Products, particularly food products for niche markets. Paper presented in the expert meeting on: *Ways to Enhance the Production and Export Capacities of Developing Countries of Agriculture and Food Products, including Niche Products, such as Environmentally Preferable Products*. UNCTAD, Geneva, 16–18 July. *Source*: www.fao.org/organicag/doc/unctad2001.htm (Accessed in August 2014).

Verma, S. and S.P. Singh (2008). Current and future status of herbal medicines. *Veterinary World* 1(11): 347–350. *Source*: www.veterinaryworld.org/2008/November/Curren t and future status of herbal medicines.pdf (Accessed in April 2015).

Villareal, M.O., S. Kume, T. Bourhim, F.Z. Bakhtaoui, K. Kashiwagi, J. Han, C. Gadhi and H. Isoda (2013). Activation of MITF by argan oil leads to the inhibition of the Tyrosinase and Dopachrome Tautomerase expressions in B16 murine melanoma cells. *Evidence-Based Complementary and Alternative Medicine*, vol. 2013, Article ID 340107, 9 pages. Hindawi Publishing Corporation. *Source*: www.ncbi.nlm.nih.gov/pmc/articles (Accessed in May 2014). Also available on: www.researchgate.net/publication (Accessed in November 2014).

Wachtel-Galor, S. and Benzie, I.F.F. (2011). An introduction to its history, usage, regulation, current trends, and research needs (Chapter 1: Herbal Medicine). In: Benzie I.F.F. and Wachtel-Galor S. (eds.), *Herbal Medicine: Biomolecular and Clinical Aspects* (2nd ed.). Boca Raton (FL), CRC Press. *Source:* www.ncbi.nlm.nih.gov/books/NBK92773/ (Accessed in April 2015).

Wade, L. (2014). Celebrated biodiversity institute faces financial crisis. *Science* 346(6216): 1440. *Source*: www.sciencemagazinedigital.org/sciencemagazine/19_december_2014 (Accessed in May 2015).

Walter, S. (2002). Certification and Benefit-Sharing Mechanisms in the Field of Non-Wood Forest Products—An Overview. Medicinal Plant Conservation—8. Newsletter of the IUCN Species Survival Commission, Medicinal Plant Specialist Group, Bonn, Germany. *Source*: www.fao.org/docrep/article/001/ab542e01.htm (Accessed in April 2014).

Walter, S., P. Vantomme, W. Killmann and F. Ndeckere (2004). Benefit sharing arrangements in the field of non wood forest products: Status and links to certification. In: Sustainable Production of Wood and Non Wood Forest Products—Proceedings of the IUFRO Division 5, Research Groups 5.11 and 5.12, Rotorua, New Zealand, 11–15 March 2003. USDA Forest Service, Oregon in cooperation with IUFRO. pp. 111–120. *Source*: www.iufro.org/ download/file/4768/4508/51100-51200-rotorua03_pdf/ (Accessed in October 2013).

Watal, J. (2001). *Intellectual Property Rights in the WTO and Developing Countries*. Oxford University Press, New Delhi.

Watson, M. and G. Gamage (1998). Status and trends in access to genetic resources and traditional knowledge in Sri Lanka. In: South and Southeast Asia Regional Workshop on Access to Genetic Resources and Traditional Knowledge, Chennai, 22–25 February. IUCN Regional Biodiversity Programme, Asia, pp. 108–118. *Source*: www.bklabs.htm (Accessed in September 2004).

Wekesa, M. (2006). What is *sui generis* System of Intellectual Property Protection? *ATPS Technopolicy Brief 13*. The African Technology Policy Studies (ATPS) Network, Nairobi, Kenya. *Source*: www.atpsnet.org/Files/technopolicy_brief_series_13.pdf (Accessed in November 2014).

WHO (2001a). TRIPS, CBD and Traditional Medicines: Concepts and Questions. Report of an ASEAN Workshop on the TRIPS Agreement and Traditional Medicine. Jakarta, 15 February. *Source*: http://apps.who.int/medicinedocs/en/d/Jh2996e/10.2.2.html (Accessed in November 2014). Also available on: www.who.int/phi/publications/category/en (Accessed in April 2014).

———— (2001b). Declaration on the TRIPS Agreement and Public Health. WTO Ministerial Conference, Fourth Session, Doha, 9–14 November. Document code: WT/MIN(01)/DEC/W/2 dated 14 November. *Source*: www.who.int/medicines/areas/policy/tripshealth.pdf (Accessed in May 2015).

———— (2002a). World Health Organization Fact Sheet No. 271. *Source*: www.who.int/medicines/organization/trm/factsheet (Accessed in September 2004).

———— (2002b). TRIPS, Intellectual Property Rights and Access to Medicines. Antiretroviral Newsletter Issue No. 8. World Health Organization Regional Office for the Western Pacific, Manila, the Philippines. *Source*: www.wpro.who.int/hiv/documents/docs/ARVnewsletter8_Jan2006update_F9F3.pdf (Accessed in February 2014).

———— (2003). WHO Paper on Traditional Medicine. Fact sheet No. 134, Revised May 2003. *Source*: www.who.int/mediacentre/factsheets/2003/fs134/en (Accessed in April 2015).

WIPO (2001). An act providing for the conservation and protection of wildlife resources and their habitats, appropriating funds therefor and for other purposes. *Source*: www.wipo.int/wipolex/en/text.jsp (Accessed in July 2014).

———— (2002). Elements of a *sui generis* System for the Protection of Traditional Knowledge. Document prepared by the Secretariat for the Fourth Session of IGC, Geneva, 9–17 December. Document code: WIPO/GRTKF/IC/4/8, September 30. *Source*: www.wipo.int/edocs/mdocs/tk/en/wipo_grtkf_ic_/wipo_grtkf_ic_4_8.pdf (Accessed in November 2014).

WIPO (2003a). Report of the Fifth Session of the IGC held at Geneva, 7–15 July. Document code: WIPO/GRTKF/IC/5/15 dated 4 August. *Source*: http://www.wipo.int/edocs/ mdocs/tk/en/wipo_grtkf_ic_5/wipo_grtkf_ic_5_15.pdf (Accessed in May 2015).

——— (2003b). Protection of Traditional Knowledge and Genetic Resources: A Bottom-up Approach to Development. *WIPO Magazine*, November–December, pp. 18–21. *Source*: www.wipo.int/export/sites/www/wipo_magazine/en/pdf/2003/wipo_ pub_121_2003_11-12.pdf (Accessed in May 2015). Previously accessed (January 2005) from: http://r0.unctad.org.

——— (2004a). Intellectual Property and Traditional Knowledge. Booklet No. 2. WIPO Publication No. 920(E). World Intellectual Property Organisation, Geneva. *Source*: www. wipo.int/edocs/pubdocs/en/tk/920/wipo_pub_920.pdf (Accessed in August 2014).

——— (2004b). Study takes critical look at benefit sharing of genetic resources and traditional knowledge. WIPO press release PR/2004/373, Geneva. *Source:* www.wipo. int/pressroom/en/prdocs/2004/wipo_pr_2004_373.html (Accessed in April 2013).

——— (2010). Document prepared by the Secretariat for the Seventeenth Session of IGC, Geneva, 6–10 December. Document code: WIPO/GRTKF/IC/17/INF/9 dated 5 November 2010. *Source*: www.wipo.int/edocs/mdocs/sct/en/wipo_grtkf_ic_17/wipo_ grtkf_ic_17_inf_9.pdf (Accessed in May 2015).

——— (2012a). The World Intellectual Property Organization Traditional Knowledge Documentation Toolkit. Consultation Draft November 1. *Source*: www.wipo.int/ export/sites/www/tk/en/resources/pdf/tk_toolkit_draft.pdf (Accessed in May 2015).

——— (2012b). Intellectual Property and Genetic Resources, Traditional Knowledge and Traditional Cultural Expressions: An Overview. WIPO Publication No. 933E, WIPO, Geneva. *Source*: www.wipo.int/edocs/pubdocs/en/tk/933/wipo_pub_933.pdf (Accessed in June 2014).

——— (2013). Matters Concerning the IGC on IP and Genetic Resources, Traditional Knowledge and Folklore. Forty-third (21st ordinary) Session of the Assemblies of Member States of WIPO (23 September to 2 October); Decision on Agenda Item 35. *Source*: www.wipo.int\... \73session_igc_mandate_1415.pdf. (Accessed in May 2014).

——— (2014a). Joint recommendation on the use of databases for the defensive protection of genetic resources and traditional knowledge associated with genetic resources. Document submitted by the Delegations of Canada, Japan, the Republic of Korea and the USA to the Twenty-Sixth Session of IGC, Geneva, 3–7 February. Document code: WIPO/GRTKF/IC/26/6; 31 January 2014. *Source*: www.wipo.int\...\wipo_grtkf_ ic_26_6.pdf (Accessed in May 2014).

——— (2014b). Report adopted by the Committee in the Twenty-Sixth Session of IGC, Geneva, 3–7 February. Document code: WIPO/GRTKF/IC/26/8; 24 March 2014. *Source*: www.wipo.int\....\wipo_grtkf_ic_27_ref_grtkf_26_8.pdf (Accessed in May 2014).

——— (2014c). The Protection of Traditional Knowledge: Draft Articles. Document prepared by the Secretariat for the Twenty-Seventh Session of IGC, Geneva, 24 March–4 April. Document code: WIPO/GRTKF/IC/27/4; 23 January 2014. *Source*: www.wipo. int\....\wipo_grtkf_ic_27_4.pdf (Accessed in May 2014).

——— (2014d). Initial Draft Report. Document prepared by the Secretariat for the Twenty-Seventh Session of IGC, Geneva, 24 March–4 April. Document code: WIPO/GRTKF/ IC/27/10 PROV; 9 May 2014. *Source*: www.wipo.int\....\wipo_grtkf_ic_27_10_prov.pdf (Accessed in May 2014).

——— (2014e). Matters concerning the IGC. Document prepared by the Secretariat for the WIPO General Assembly, Forty-Sixth (25th Extraordinary) Session, Geneva, 22–30 September. Document code: WO/GA/46/6; 22 July 2014. *Source*: www.wipo.int\....\ wo_ga_46_6.pdf (Accessed in September 2014).

WIPO (2014f). Documenting Traditional Medical Knowledge. Report prepared by Ryan Abbott. *Source*: www.wipo.int/export/sites/www/tk/en/resources/pdf/medical_tk.pdf (Accessed in April 2014).

———— (undated). Leveraging Economic Growth through Benefit Sharing. WIPO programme activities. *Source*: www.wipo.int/ipadvantage/en/details (Accessed in May 2014).

WTO (2000). Protection of Biodiversity and Traditional Knowledge—The Indian Experience. Submission by India in the Committee on Trade and Environment, Council for TRIPS of the WTO (WT/CTE/W/156/IP/C/W/198 dated 14 July). *Source*: http://commerce.nic.in/trade/international_trade_papers_nextDetail.asp?id=47 (Accessed in September 2014).

———— (2001). Doha Ministerial Declaration 2001. WTO Ministerial Conference, Fourth Session, Doha, 9–14 November. Document code: WT/MIN(01)/DEC/1 dated 20 November. *Source*: www.wto.org/english/thewto_e/minist_e/min01_e/mindecl_e.pdf (Accessed in September 2014).

Wynberg, R. (2004). Rhetoric, realism and benefit sharing: Use of traditional knowledge of *Hoodia* species in the development of an appetite suppressant. *Journal of World Intellectual Property* 7(6): 851–876. *Source*: www.biowatch.org.za/pubs/wjip.html (Accessed in August 2013).

Wynberg, R. and R. Chennells (2009). Green diamonds of the South: An overview of the San-Hoodia case. In: R. Wynberg, D. Schroeder and R. Chennells (eds.), *Indigenous Peoples, Consent and Benefit Sharing: Lessons from the San-Hoodia Case*. Springer Netherlands, pp. 89–124. *Source*: www.msu.ac.zw/elearning/material (Accessed in August 2013).

Yadav, S.K. and G.C. Mishra (2013). Biodiversity management open avenues for bioprospecting. *International Journal of Agriculture and Food Science Technology* 4(6): 635–642. *Source*: www.ripublication.com/ijafst_spl/ijafstv4n6spl_17.pdf (Accessed in May 2014).

Zuidema, P.A., J. Sayer and W. Dijkman (2005). Forest fragmentation and biodiversity: The case for intermediate-sized conservation areas (Chapter 15). In: Jeffrey Sayer (ed.), *The Earthscan Reader in Forestry and Development*. Earthscan, London and Sterling VA (USA), pp. 306–322.

Index

access and benefit sharing (ABS)
  best course of action for
    mechanism of benefit sharing,
      salient features of, 273–275
  mechanism for, 208f
  new scienftific-cum-legal disci-
    pline, 269–271, 272f
  professional, 271, 273
  regulations and biodiversity laws,
    110
    Costa Rica, 111–112, 113
    Philippines, role of, 114
    Provisional Measure, 112
    Wildlife Resources
      Conservation and Protection
      Act, 2001, 114
access and benefit sharing (ABS)
    regulations
  Biological Diversity Act of 2002,
    109
  India's national biodiversity
    authority, 120–125
  India's Patents Act of 1970,
    108–109
  objective of, 109
  North–South divide and, 110
  patent filing in India, 118–120
  State Biodiversity Board (SBB), 109
  WTO, backdrop of implications,
    108
Act on the Protection of Access
    to Peruvian Biological

Diversity and the Collective
    Knowledge of Indigenous
    Peoples, 114
  functions, 115
Agreement on TRIPS, 81, 284
  Article 27, dialogue on, 95–99
  Article 29, 100b
  Article 7, 82
  Articles 29, 285
  Articles 7 and 8, 285–286
  limitations of, 93–95
  review of, 103–107
Aguaruna People of Peru and Maya
    People of Mexico, 168
Argan Oil Commercialization, 195
  argan ecosystem, 196–197
  argan oil, 197–199
  argan tree, 196
  benefit sharing arrangements,
    200–201
  commentary on, 202–203
  corporate social responsibility, 201
    aspects of, 202
  options available, 203–205
  women's cooperatives, 199–200

benefit sharing
  bioprospecting partnership agree-
    ments, formulation, 231–235
  groups, defining of, 214–217
  ICBG projects, area and size
    delineation of, 211–214

mechanism for, 208–209
  bioprospecting partnership elements, 210
  representative kind of, 210–211
partnerships and time frame, framework for, 227–228
resource, inventory of, 228–230
stakeholders, identification and analysis of, 217–227
participatory learning and action (PLA), 225
SWOT analysis, 223–224
benefit sharing arrangements
bioprospecting projects in
  types of benefits, 235–237
other types, 240–242
socio-economic and infrastructure development works, 246–248
trust fund
  constitution of, 243–246
  mechanism of, 243, 244f
types of, resource providers, 235
  advance payments, 236
  capacity building, 237–238
  equipment, training and infrastructure, 236
  material-individual rewards, 237
  milestone payments, 236
  royalty, 236
  sustainability, 238–240
  upfront payments, 236
benefit sharing with Kani people, 178–180
  commentary on attempts made, 186–188
  conflicts, 182–185
  drug development
    deed of trust states, objectives, 182
    WIPO-UNEP study, 180–181
biodiversity
  'thinking globally, acting locally' cases, 5

Agreements on TRIPS and CBD symbiosis between, 9
Article 27, 4
Article 41, 8
BDCP, 2
CBD, 1
  commercial use of, 1–3
    international instruments support, 278–289
  conservation of, 32–41
    BHS, 36
    Biological Diversity Act, 36–37
    competing paradigms for forests, 35–36
    forest fragmentation, 35
    Task Force, 38–39
  COP, 7
  CSR
    approach, 3
  Doha Declaration, 7
  equitable benefit sharing, 7
  FIRD-TM, 2
  INBio, 1–2
    example of, 2
  IPRs, 1
  JNTBGRI, 3
  Morocco, 3–4
  Nagoya Protocol, 7–8
  new prospecting project
    precautions of, 276–277
  ownership issue of natural resources, 6–7
  scanty data of trade in, 19–22
    CIS countries, 19
    gross domestic product (GDP) and, 19
    UNDP, 22
  usufructs, 5
Biodiversity Heritage Sites (BHS)
  features of, 36
biodiversity information dissemination programme, 151
Biodiversity Law of South Africa, 116
  functions, 115

Regulations, 118
South Africa's Biodiversity Act, 117
Biodiversity Laws of Peru, 48b
biodiversity prospecting programme
INBio-Merck partnership, 153–154
monetary benefits, 152
non-monetary benefits, 152
biodiversity-rich tropical countries
recent developments in, 289–296
biological diversity
functional tool, 263
issues to resolve, 263
status of, 15–19
Biological Diversity Act, 36–37
biological resources
trade, salient features of, 266–267
biopiracy, 51
definition of, 52
Erosion, Technology and
Concentration (ETC) Group,
52–53
bioprospecting partnership agreements
formulation of, 231–235
JNTBGRI, 234
MOU, 231
Bioresources Development and
Conservation Programme
(BDCP), 2
Bonn Guidelines, 264–265b

cartagena protocol on biosafety, 264b
commercial use of biodiversity
benefit sharing arrangements,
analysis of, 141–146
case studies of, 144–145t
INBio-Merck Collaboration,
146–158
commonwealth of independent states
(CIS), 19
components of biodiversity
biodiversity research and bioprospecting, 26–32
commercial uses of

investigations into, 15
international trade, 14, 22–26
biological resources, 25
MAPs, 11
components of, 12
natural resources
commercialization of, 14
NWFPs
definitions of, 13
sustainable use of, 32–41
trades of
NWFPs, 11–12
traditional knowledge, 26–32
WHO, estimation of, 12
conference of the parties (COP), 7
convention on biological diversity
(CBD), 1, 86–88
limitations of, 88–90
mandate of, 99–100
corporate social responsibility (CSR)
approach, 3

Doha Ministerial Declaration 2001,
281b

Forest People's Fund (FPE), 164–165
forging bioprospecting partnership
contracts, 206–208
Fund for Integrated Rural
Development and
Traditional Medicine
(FIRD-TM), 2, 170–173
benefit sharing arrangements
experiment of, 176
guidelines, 174–175
principles, 175
commentary on fund, 177–178
constitution of fund
advisory body, 174
board of management, 174
board of trustees, 173

General Agreement on Tariffs and
Trade (GATT), 49

INBio-BTG Agreement, 155
INBio-DIVERSA Agreement, 155
INBio-INDENA Agreement, 155
INBio-Merck Collaboration,
    146–148
  benefit sharing arrangements,
    149–150
  commentary on, 156–158
  INBio's obligations, 148
  Merck's obligations, 148–149
  national biodiversity inventory,
    150–151
  programme
    biodiversity information dis-
      semination, 151
    biodiversity information man-
      agement, 151
    biodiversity prospecting,
      152–154
    lateral collaborative agreements,
      154–155
India's Biological Diversity Act
  Agreement on TRIPS and CBD
    comparison between, 129–140,
      132–139t
  benefit sharing mechanism in, 125,
    128
  concerns in ABS, 128t
  manners to follow, 126
  provisions of, 127
India's Biological Diversity Act of
    2002, 41, 69
  access to genetic resources,
    286–287
  enforcement of IPRs, 286
  mechanism for benefit sharing, 287
  protection of TK, 286
India's national biodiversity authority,
    120–125
  applications received by, status of,
    124t
  Biological Diversity Act of 2002,
    120
  novelty in, 123

Scheduled Tribes and Other
    Traditional Forest Dwellers,
    120–121
India's Patents Act of 1970
  objective of, 109
Instituto Nacional de Biodiversidad
    or National Institute of
    Biodiversity (INBio), 1–2
  example of, 2
Intellectual Property and Genetic
    Resources, Traditional
    Knowledge and Folklore
    (IGC), 84, 279
  reiterations of, 290
intellectual property rights (IPRs), 1
International Cooperative Biodiversity
    Groups (ICBG)
  programme
    advance payments, 163
    equipment, training and infra-
      structure, 163
    priority research areas, 163
    royalties, 163
  projects
    area and size, delineation,
      211–214

Jawaharlal Nehru Tropical Botanic
    Garden and Research
    Institute (JNTBGRI), 3

lateral collaborative agreements, 154
  INBio-BTG Agreement, 155
  INBio-DIVERSA Agreement, 155
  INBio-INDENA Agreement, 155

material transfer agreement (MTA)
  certificate of origin and, 58–59
    mutually agreed terms (MAT),
      58
  Patents Act of India, 59
medicinal and aromatic plants
    (MAPs), 11
  components of, 12

Nagoya Protocol, 7, 282–283b
    access and benefit sharing, 295b
    pillars of
        access, 8
        benefit sharing, 8
        compliance, 8
        ongoing deliberations, 8
non-wood forest products (NWFPs),
        11–12
North–South divide, 90–93

one-click database search system,
        291

participatory learning and action
        (PLA), 225
Patents Act of India, 59
Peru Project and Maya ICBG Project,
        167–170
Prior Informed Consent (PIC)
    requirement, 57

registration system
    Biological Diversity Act of 2002, 69
    idea behind, 67–68
    regulations, 118
Research and Information System
        (RIS)
    developing countries for, 55
resource provider countries
    bargaining power of, 249
        weaknesses of the source coun-
            tries, 250–252
    recommendation, 268–269

San Hoodia case of benefit sharing,
        188–190
    commentary on, 193–195
    negotiations for, 190–192
    San Hoodia Benefit Sharing Trust,
        192–193
source countries
    genetic resources of, 252–255
    intellectual property, 261–262

    prior informed consent,
        255–257
    tough negotiations, need for,
        257–261
South Africa's Biodiversity Act, 117
strengths, weaknesses, opportunities
        and threats (SWOT) analysis,
        223–224
Suriname ICBG Project, 159–160
    benefit sharing arrangements
        contracts, preparation of,
            162–164
    commentary on, 165–167
    FPE, 162–164
    Peru Project and Maya ICBG
        Project, 167–170
    programme design, 161

Task Force, 38–39
traditional knowledge (TK)
    Agreement on TRIPS, 50–51
    documentation of, aspect, 62–65
        digital library, 66–67
        people's biodiversity registers,
            65–66
        registration system, 67–69
    IPR regime in, 42–46
        benefit claimers, 44
        definition by Johnson (1992),
            42–43
        significance of, 46
    mode of protection of, 46–51
        GATT, 49
    options for protection of
        MTA, 58–59
        PIC, requirement of, 57
        prior art, 56–57
        Sui Generis Legislation,
            59–61
    protection of, 252–255
        biopiracy and, 51–56
    public domain, imbroglio of, 69,
        71–2
        third world network (TWN), 70

United Nations Development
  Programme (UNDP), 22

Wildlife Resources Conservation and
  Protection Act, 2001, 114
World Trade Organization (WTO)
  and North–South divide
  advent of, 78–81
  Agreement on TRIPS, 81–82

benefit sharing arrangements,
  anomalies in, 74–78
CBD, 86–88
implications of, 73–74
limitations of, 88–90
varieties of genetic material
  and life forms, patenting of,
  84–86
arrival of, 284

About the Author

Shivendu K. Srivastava served in the Indian Forest Service (Madhya Pradesh cadre) till recently. Prior to his last assignment as Project Director, livelihoods development project of UNDP in Bhopal, he was associated with Dehradun-based Forest Survey of India (FSI), an institution responsible for reporting on the status of forest cover of the country, under the Ministry of Environment, Forests and Climate Change. He was involved in forestry research (1997–2002) at Forest Research Institute, Dehradun, where, targeting forest-based enterprises, he took up far-reaching extension work.

His varied experiences include catching live a man-eater tigress and its cub (1991), introducing the first 'Biodiversity Conservation Working Circle' in Madhya Pradesh (probably even in India) and calculating the rates of timber bought from private growers by graphical method. He has been a member of the central editorial board of *Indian Forester* (a peer-reviewed international journal since 1875). He is fond of teaching, and he is a poet too.